THE CHEMISTRY & PHARMACY

OF BOTANICAL MEDICINES

Charles Dickson

The Chemistry and Pharmacy of Botanical Medicines

ISBN: 978-0-8206-0239-4
eBook: 978-0-8206-0418-3

First Edition:
Chemical Publishing Company, Inc. - 2016

Chemical Publishing Company:
www.chemical-publishing.com

Printed in the United States of America

THE AUTHOR

Dr. Charles Dickson is a retired chemistry professor. He did graduate study at the University Florida under a National Institutes of Health Fellowship in Pharmacology. He has taught at numerous colleges in Florida and North Carolina. He has authored two general chemistry laboratory manuals and two previous books in pharmaceutical chemistry.

He lives in Hickory, North Carolina and continues to teach part-time at the Catawba Valley Community College in that city.

PREFACE

A celebrated 16th century English divine, William Cole, said of herbs, "It is a subject as ancient as Creation, yet more ancient than the sun or moon or stars, they being created on the fourth day, whereas plants were the third." Today's evolving interest in ecology and the preservation of nature has led us to reconsider the importance of the plant world as it relates to our health.

To be sure, plants have always occupied a prominent part in the healing arts throughout history and to our present day. The contemporary environmental scientist and author, G. Tyler Miller, reminds us that one-fourth of all the medicines we see on our modern pharmacy shelf had their origins in plant use. Of course we don't sell willow bark over the counter now—we simply synthesize the aspirin tablet from it.

This text, *The Chemistry and Pharmacy of Botanical Medicines*, is organized by major therapeutic category. Hopefully it brings together the achievements of three major disciplines to help us understand and appreciate the importance of some select medicinal plants.

TABLE OF CONTENTS

TABLE OF APPARATUS ILLUSTRATIONS

ANALGESICS

PHARMACY – Analgesics are agents which relieve pain by acting centrally to elevate the pain threshold without disturbing consciousness or altering sensory modalities. It has been concluded that pain cannot be defined except one defines it introspectively for himself. Many drugs which relieve pain cannot be classified as analgesics since some produce hiatus of consciousness and others block peripheral nerve fibers. There are many synthetic analgesics on the market. Some botanical ones include the following:

OPIUM

BOTANY – The Opium poppy, called *Papaver somniferum*, is an annual that grows 30-150 cm high. It is one-stemmed, blue-gray frosted plant, which produces a white milky latex. The plant originated in western Asia but is now cultivated worldwide commercially. The medicinal part is the latex extracted from the seed capsule.

CHEMISTRY – Compounds of the opium poppy latex include isoquinoline alkaloids (20-30%), codeine (0.2- 3.5%) paperine (0.5-3.0%). The most important alkaloids which have been isolated from opium are morphine and codeine.

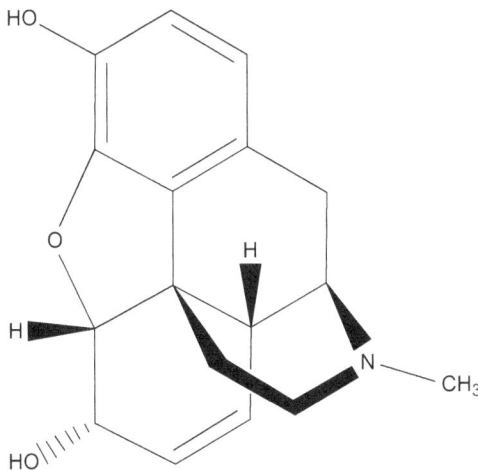

Figure 1. Morphine molecule

Triturate 10 gm of opium powder with sufficient calcium chloride to form a thin paste and extract with hot water. This converts the morphines and other alkaloids into their respective tydrochlorides while the acids which were combined are precipitated as calcium meconate and calcium sulfate. The insoluble matter is separated by filtration, and to prevent oxidation, sodium sulfite is added to the filtrate which is then concentrated in vacuo to the consistency of a thin syrup. Sodium acetate solution is added in order to precipitate narcotine and papaverine which are removed by filtration. A small proportion of ethanol is added to the filtrate. The morphine is precipitated by the addition of calcium oxide in the presence of ammonium chloride. After allowing to stand for a while, the morphine is filtered off. The crude drug is washed with benzene to remove traces of codeine, then mixed with boiling water and neutralized with hydrochloric acid. Atmospheric oxidation should be prevented by covering with a layer of petroleum jelly. Morphine hydrochloride crystallizes out on cooling and is re-crystallized from water.

Tests for Morphine: Add a drop of nitric acid to the solid morphine or one of its salts. An orange-red color is produced as evidence of the presence of morphine.

Morphine is also able to reduce ferricyanide to a ferrocyanide. The test is conducted by adding a drop of potassium ferricyanide (which has been mixed with a smaller drop of ferric chloride) to a liquid containing a small amount of morphine. The presence of morphine is evidenced by the appearance of a bluish-green color.

Another reaction of morphine is conducted by adding a crystal of sodium nitrite to an acid solution of the alkaloid followed by the addition of an excessive amount of ammonium hydroxide. This reaction yields a brownish-yellow color.

BELLADONNA

BOTANY – *Atropa belladonna* is a perennial herbaceous plant, 1-2 meters high with a many-headed cylindrical rhizome. The woody stem is erect, branched, bluntly-angular, and hairy. It grows throughout western, central, and southern Europe and also the Balkans, Southeast Asia, Iran, northern Africa as well as Denmark, Sweden, and Ireland.

CHEMISTRY – Its chief compound is an alkaloid called hyoscyamine, which can be transformed into atropine, scopolamine, and tropine. The assay of Belladonna for hyoscyamine is performed for purposes of standardization proof of purity, commercial evaluation, and pharmacology purposes. The amount of alkaloids that occur in crude drugs are subject to considerable variation in different samples of the same drug. The variations may be caused by:

☐ The age of the plant when it is collected
☐ The season of the year when the drug is harvested

☐ The soil and climate in which the drug is grown

☐ Conditions under which the drug is collected and stored

<u>Alkaloidal determination in Belladonna</u>: Place 10 gm of belladonna powder in a dry flask and macerate with 10 ml of ethanol and 20 ml of ether for 10 minutes, then add 5 ml of ammonium hydroxide to render the solution alkaline and liberate the alkaloidal bases from their salts. After allowing to stand with frequent shaking for an hour, transfer the contents of the flask to a small percolator. Percolate with an ether alcohol mixture until the alkaloids are extracted. In order to ascertain when complete extraction has taken place, a few drops of the solvent is collected and placed on a watch glass. When the solvent evaporates, treat the residue with a few drops of dilute hydrochloric acid and a few drops of Mayer's Reagent. If extraction is incomplete, a cream-colored precipitate will appear.

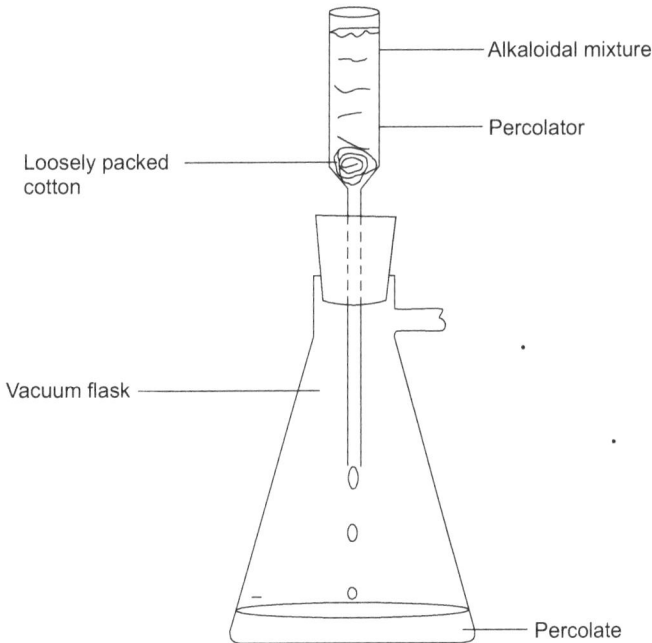

Figure 2. Percolator

Once extraction is complete the percolate may now be transferred to a separatory funnel and shaken with an excess of dilute hydrochloric acid. After separation the lower level is drawn off into another separatory funnel and the ethereal layer twice more extracted with small portions of dilute hydrochloric acid mixed with a little alcohol. The acid solution of the alkaloids is now freed from traces of chlorophyll by adding 10 ml of chloroform, allowing it to separate and drawing off the chloroformic liquid into another separatory funnel and shaken with a little dilute acid to remove traces of alkaloids.

The acidic solution is then rendered alkaline with 10 ml of ammonium hydroxide. This liberates the alkaloids which can then be extracted with several successive portions of chloroform. Two ml of ethanol should be added to the residue the evaporated off a water bath and the residual alkaloid dried at 100 °C. The residue is now dissolved in 10 ml of .02 M hydrochloric acid and the excess titrated with .02 M sodium hydroxide using methyl red as an indicator. Each ml of .02 M hydrochloric acid used is equivalent to 0.0055784 gm of hyoscyamine.

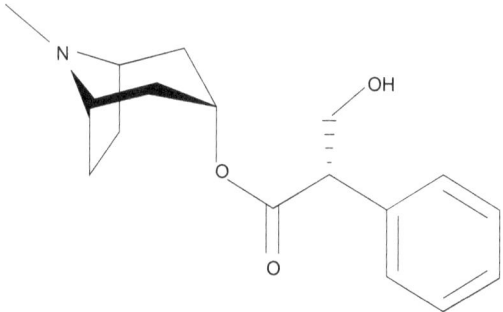

Figure 3. Hyoscyamine molecule

GELSEMIUM

BOTANY – *Gelsemium sempervirens* is a perennial evergreen vine on a tortuous, smooth root with a thin bark and woody center showing broad medullary rays. There are yellow flowers, strongly perfumed 2.5-4.cm long and funnel-shaped. The medicinal parts are the dried rhizome and the roots. The plant is indigenous to North America along the coast from Virginia to Florida and Mexico

CHEMISTRY – The main alkaloid is gelsemine, but there are also other compounds including gelsimicin, gelsidin, scopoletin, gelseviren, and semperviron plus various hydroxycourmarins.

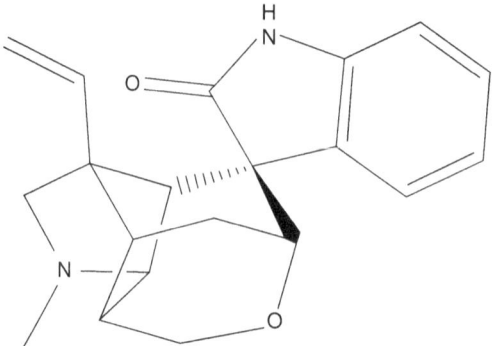

Figure 4. Gelesimine molecule

<u>Test</u>: Add a few drops of nitric acid to a small amount of gelsemine. Note the color change from red to dark green.

ACONITE

BOTANY – *Aconite napellus* is a 0.5 to 1.5 meter high shrub with a tuberous thickened fleshy root and an erect, rigid, undivided stem. The racem axis and petioles are glabrous or hairy and the leaves are dark green, glossy above and lighter beneath. Known popularly as Monkshood, the plant is common to the Alps and Carpathians and in most mountainous regions of Europe. It is found as far north as Sweden and to England and Portugal in the west.

CHEMISTRY – Compounds are mainly terpene alkaloids including aconitine, mesaconitine, and hypaconitine.

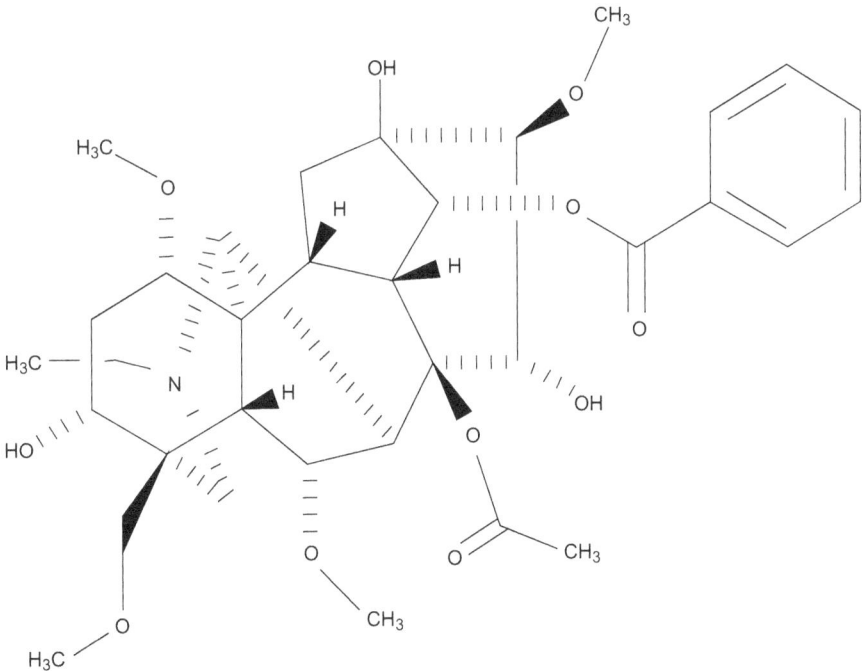

Figure 5. Aconitine molecule

<u>Preparing a tincture</u>: Mascerate 100 gm of the drug in a 3:1 ethanol:water solution, then percolate it for 24 hours at a moderate rate. Then add sufficient hydrochloric acid to produce a pH of 3.0. Assay a portion of the percolate and adjust volume of remaining fluid to a 3.0 pH.

FEVERFEW

BOTANY – The plant is a strongly aromatic perennial with leaves that are pin-natisect to pinnatifid and yellowish-green. The basal and lower cauline leaves are more or less ovate with 3-7 oblong-elliptical to ovate segments which are subpinnately divided and crenate or entire-marginal. Known botanically as *Tanacetum parthenium* the plant originated in southeastern

Europe but is now found all over the continent, Australia, and North America.

CHEMISTRY – The chief constituents of Feverfew are volatile oils, including camphor, trans-chrysanthylacetet, and a variety of terpenes.

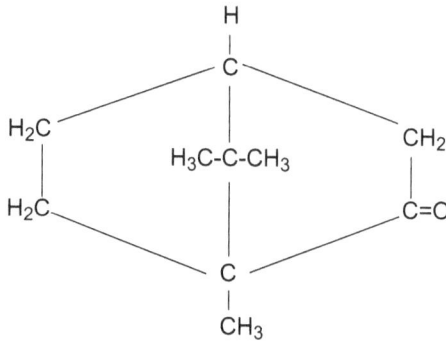

Figure 6. Camphor molecule

Assay for camphor: Place 2 ml of camphor spirit in a pressure bottle containing 50 ml of dinitrophenylhydrazine. Close the pressure bottle, immerse in a beaker of water and heat on a stem bath for 4 hours. During the heating maintain a temperature of 75 °C. Cool the pressure bottle and contents to room temperature and transform the contents to a beaker while adding 100 ml of 0.1 M sulfuric acid. Allow to stand not less than 12 hours at room temperature. Transfer the precipitate to a previously dried and weighed filtering crucible and wash with 100 ml of 0.1 M sulfuric acid followed by 75 ml of cold water. Continue the solution until the excess water is removed. Dry the crucible and precipitate at 80 °C for two hours. Each gm of camphor dinitrophenylhydrazine is equivalent to 458 gm of camphor ($C_{10}H_{16}O$).

ADDITIONAL botanicals with analgesic actions include catechu, calendula, berberis, black cohosh, turmeric, wild yam, St. John's Wort, and veratrum.

ANTHELMINTICS

PHARMACY – Anthelmintics are drugs which drive parasites from the intestinal tract. These parasites belong either to the Namathelminthes (round worms) or Platyhelminthes (flat worms) phyla. Anthelmintics are most effective when there is no material in the intestinal tract. As parasitic worms are harmful to the host for a number of reasons, it is desirable to eliminate them as soon possible. Such organisms deprive the host of food. They may injure organs by obstructing ducts, or they may secrete substance toxic to the host, be it human or animal. Even though the host may be symptom-free, it is recommended that the parasites be eradicated as soon as they have been discovered.

SANTONIN

BOTANY – Santonin is obtained from the unexpanded flower heads of *Artemista cina*, also known as Wormwood. It is an evergreen perennial semi-shrub 30-60 cm high with many slim, sprouting stems and is indigenous to Iran, Turkestan, and the Kirghizia Steppes around Bukhara. The medicinal part is the flower.

CHEMISTRY – Santonin compounds consist of sequiterpenes lactones including alpha and beta santonin and artemisin.

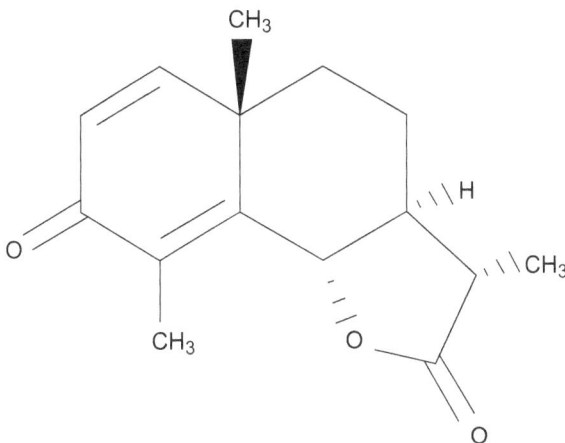

Figure 7. Santonin molecule

Test: Heat about 0.2 gm of santonin in 2 ml of alcoholic potassium hydroxide. A red color appears. Shake .01 gm of santonin with a cool mixture of 1 ml each of sulfuric acid and distilled water. Now heat to 100 °C and add a few drops of ferric chloride TS. This reaction yields a violet color. Both are tests for the presence of santonin.

CHENOPODIUM

BOTANY – Chenopodium is a plant about 1 meter high with branched, reddish stems covered in alternate linear to lanceolate leaves and grown mainly in Mexico and South America. The medicinal parts are the seeds including the flowers.

CHEMISTRY – The main chemical constituent of chenopodium is Ascaridol, an organic peroxide.

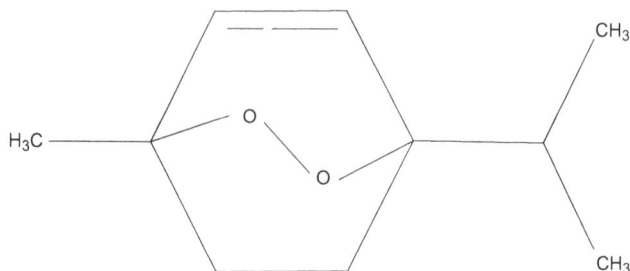

Figure 8. Ascaridol molecule

Test: Organic peroxides tend to explode when heated with organic acids. Place 2 ml of ascaridol in a small beaker and add 2 ml of acetic acid. Heat carefully and note the results.

QUASSIA

BOTANY – Quassia is a tree known botanically as *Picrasma excelsa*. It is a tree which grows to 15-30 meters high with a diameter of one meter. The bark is smooth and gray with alternate leaves that are odd-pinnate. The tree grows mainly in Jamaica. Its medicinal part is the wood of the trunk and branches.

CHEMISTRY – Quassia contains decanor-triterpenes and Quasinoids including nigakilactone D, neoquassin, and 18-hydroxy quassin beta carboline alkaloids.

Test: Prepare tincture of Quassia by mixing 30 gm of the bark with 40 ml of ethanol and 80 ml of distilled water and percolate for two hours. Pour the percolate out as the tincture.

CASTOR OIL

BOTANY – *Ricinus comunis* is the botanical name for Castor Oil. It is an annual plant in central Europe which has a taproot and lateral roots near the surface. The stem has alternate palmate and reddish, simple long-petioled leaves arranged in a spiral. The leaf blade peltate 10-60 cm long and usually divided into palmate ovate-oblong or lanceolate lobes. It thrives in most temperate latitudes.

CHEMISTRY – Castor Oil is 42-55% fatty oils with proteins, pyridine alkaloids, triglycerides, and tocopherols also present.

Iodine test for the purity of the oil: The iodine test of an oil is the number of grams of iodine absorbed under prescribed conditions by 100 gm of the oil or fat. The Hanus Method calls for introducing 250 mg of the castor oil into a glass-stoppered bottle of 250 ml. Dissolve the oil in 10 ml of chloroform and 25 ml of iodobromide TS. Stopper the flask securely and allow it to stand 30 minutes protected from light. Add 30 ml of potassium iodide TS and 100 ml of distilled water. Titrate the liberated iodine with 0.1 N sodium thiosulfate. Shake thoroughly after each addition of the thiosulfate or agitate with a magnetic stirrer. When the iodine color becomes pale add 1 ml of starch TS and continue to titrate with 0.1 N sodium thiosulfate until the blue color is discharged. Next run a blank through containing no oil and repeat the procedure. Titrate in the same manner.

The difference between the number of ml of thiosulfate consumed by the blank test and the actual test multiplied by 1.269 and divided by the weight in grams of the sample tested is the iodine number of the oil. For Castor Oil your calculations should be in the 83-89 range.

Determine the acid value of the fatty acids in Castor Oil by dissolving l0 gm of the oil in a 50 ml mixture of ethanol and diethyl ether. You may have to heat it in a reflux condenser to get everything dissolved. Now add 1 ml of phenolphthalein TS and titrate with 0.1 M sodium hydroxide until the solution becomes faintly pink after shaking for 30 seconds. Calculate the acid value by measuring the ammonium 0.1 M alkali required to neutralize the 10 gm sample.

ADDITIONAL botanicals with anthelmintic properties include calotropia, larkspur, black hellebore, yellow lupin, groundstet, and yew.

ANTI-INFLAMMATORIES

PHARMACY – The term, anti-inflammatory is not a specific class of drugs but rather a general mode of action that may occur in various areas of the body. Anti-inflammatories may include drugs which are anti-arthritic, anti-rheumatic, anti-spasmodic, anti-asthmatic, muscle relaxants, and others. They may impede inflammation of various areas such as the joints in the skeletal system or the respiratory system, or numerous organs including the skin.

ARNICA

BOTANY – *Arnica montana* is a 20-50 cm herbaceous plant with a 0.5 cm thick by 10 cm long usually unbranched, 3-sectioned, sympodial brownish rhizome. The leaves are in basal rosettes and the terminal composite flower is located in the leaf axis of the upper pair of 1eaves. The plant is found in Europe from Scandinavia to southern Europe and also in southern Russia and central Asia. The medicinal part is the flower.

CHEMISTRY – Constituents include sequiterpene lactines of the pseudo-guaianolide type plus, various volatile oils such as thymol and thymol esters and polyenes, hydroxycoumarine, and helenaline derivatives.

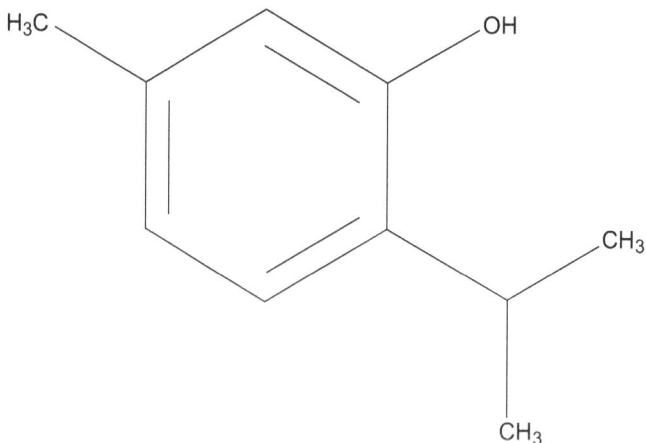

Figure 9. Thymol molecule

To make tincture of Arnica, a widely-used skin anti-inflammatory, place 25 gm of dried arnica flowers in 300 ml of a 3:1 ethanol-water mixture. Then macerate the drug during a 24-hour period. After this period percolate this mixture at a moderate rate. Many tinctures are made in similar fashion to this procedure as a tincture is always defined as an alcoholic solution of the botanical.

WHITE WILLOW

BOTANY – The White Willow, known botanically as *Salix alba,* is a 6-18 meter high tree or bush with fissured gray bark and tough sometimes egg-yellow or red-yellow supple branches. The leaves are short petioled, lanceolate, acuminate becoming cuneate at the base. The tree is indigenous to central and southern Europe. The medicinal part is the bark usually considered to be the natural origin of the modern aspirin.

CHEMISTRY – The major compounds are glucosides and esters yielding salicylic acid and salicin derivatives acylated to the glucose residue including, among others, the compounds fragilin and populin.

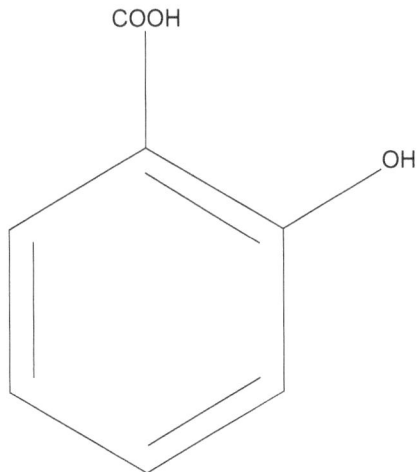

COOH

OH

Figure 10. Salicylic Acid molecule

Salicylic Acid: One test for the presence of salicylate consists of adding some drops of ferric chloride TS to a solution of salicylic acid or corresponding salicylate. The reaction gives a deep violet color.

Another test for the presence of salicylates: Heat 1 gm of salicylic acid in 3 ml of distilled water until the acid is dissolved. Thoroughly cool the solution by immersing in an ice bath. Filter the solution and. add to 15 ml of the filtrate, 2 drops of hydrochloric acid and 5 drops of barium chloride TS. No turbidity develops as a positive test for the presence of salicylates.

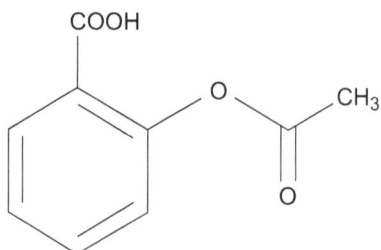

Figure 11. Acetylsalicylic Acid molecule

BUCHU

BOTANY – *Barosma betulina* is a small shrub with light green to yellowish leaves. The leaves are 12-20 mm in length, opposite, rigid, and coriaoeous. They are rhomboid or obovate short-petioled with oil gland on each indentation. The medicinal parts are the leaves. The plant is indigenous to South Africa.

CHEMISTRY- The chief component is diosphenol and psi-diosphenol (known as buccocamphor) and some flavonoids. The presence of phenol can be shown by placing 0.5 gm of buchu leaf powder and 0.5 ml of phenylhenyl isocyanate in a 25 ml flask and add 1 drop of pyridine. Stopper the flash loosely with cotton and heat for 15 minutes. If separation of the phenolic does not occur, you can cool and induce crystallization by scratching the walls.

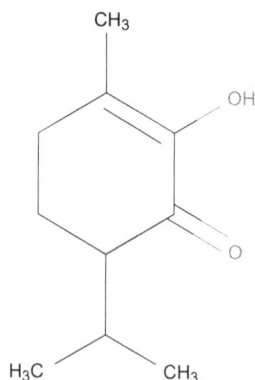

Figure 12. Diosphenol molecule

COMFREY

BOTANY – Comfrey, or *Symphytum officinale*, grows from 30-120 cm high. The root is fusiform, branched, black on the outside, white on the inside The stem is erect and stiff-haired. The leaves are wrinkly and roughly pubescent. The flowers are dull purple or violet. The plant is indigenous to Europe and temperate Asia and is naturalized in the United States.

CHEMISTRY - Compounds include allantoin, mucilage, triterpenes saponins, silicic acid, and some pyrolizidine alkaloids.

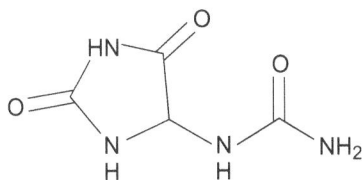

Figure 13. Allantoin molecule

Test: Allantoin as a substituted urea will show a positive Biuret reaction. Place 0.5 gm in a small test tube and add a few drops of Biuret reagent TS. The result of the test should be a color reaction ranging from violet to reddish-brown.

GUAIAC

BOTANY – Officially known as *Guiacum officinale*, the Guaiac tree is an evergreen which grows up to 13 meters high with a greenish-brown, almost always twisted trunk with furrowed bark. The leaves are short petioled, coriaceous di to tri-pinnate. The medicinal parts are the wood and the various preparations of the resin of the heartwood. The tree grows in Florida, on the Antilles, in Guyana, Venezuela, and Columbia.

CHEMISTRY – Constituents include the terpene, saponins including aglycone and oleanolic acid, plus resins and volatile oils.

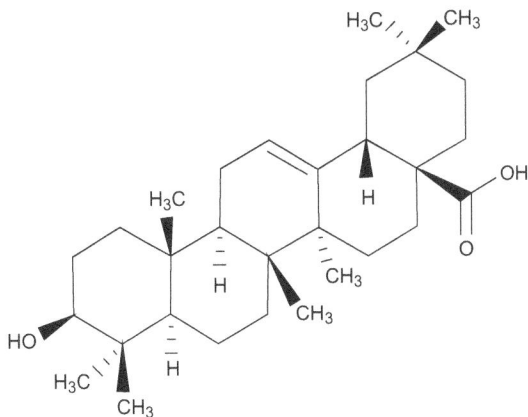

Figure 14. Oleanolic Acid molecule

Test: Add 1 drop of ferric chloride TS to 5 ml of an alcoholic solution of Guaiac (1%). A blue color should be produced which gradually changes to green, finally becoming greenish-yellow.

A second test is to place 5 ml of an alcoholic solution of Guaiac (1%) and add 5 ml of water to which 20 mg of lead peroxide has been added. Filter and boil a portion of the filtrate. The color disappears but may be restored by the addition of lead peroxide.

ADDITIONAL botanicals which possess anti-inflammatory properties include Quince, Sage, Chamomile, Monkshood, Calamus, Spikenard, Borage, Birch, Caltropis, Cumin, Male Fern, Eucalyptus, Devils Claw, Winter Cherry, Smart weed, Wafer Ash, Buttercup, Black Currant, Soapwort, Vervain, and Prickly Ash.

ANTI-MALARIALS

PHARMACY – Anti-malarials constitute a specific group of drugs which are effective, to varying degrees against several species, the protozoa *Plasmodium vivax* and *Plasmodium falciparum*. They have complex life cycles involving both the anopheles mosquito and the erythrocytes of the human host. The ideal anti-malarial should not only eradicate the microzoan from the blood but from the tissues as well.

QUININE

BOTANY – The *Cinchona pubescens* plant is an evergreen tree, sometimes a bush, which grows for 5-15 meters high with a dense crown. The branches are at right angles to the trunk. The young branches are usually pubescent. The bark occurs in quills or flat pieces up to 30 cm long and 3-6 mm thick. The medicinal part is the dried bark of 6-8 year old trees. Cinchona is indigenous to mountainous regions of the tropical United States, but it is also cultivated elsewhere.

CHEMISTRY – Quinoline alkaloids are the main components including quinine, quinidine, chinchonine, and some tannins.

To isolate the bark: Bark is treated with 30 parts of its weight by calcium hydroxide and an aqueous solution of 1.0 M sodium hydroxide. The mixture is extracted in a steam bath rotating ball mill with hot heavy petroleum which dissolves the alkaloids. After 3 hours of constant agitation the mixture if allowed to settle and the petroleum drawn off. This should be followed with more extractions and the petroleum liquor shaken with. Hot dilute sulfuric acid to form the neutral sulfates of the alkaloids. The oil should then be separated while hot and the neutral aqueous liquid cooled. Quinine sulfate separate and can be subsequently purified by recrystallization from water. Quinine alkaloid may be obtained from the sulfate by dissolving it in a large volume of water containing a little sulfuric acid and pouring the liquor into a dilute aqueous sodium carbonate solution. The precipitate quinine can then be filtered and washed.

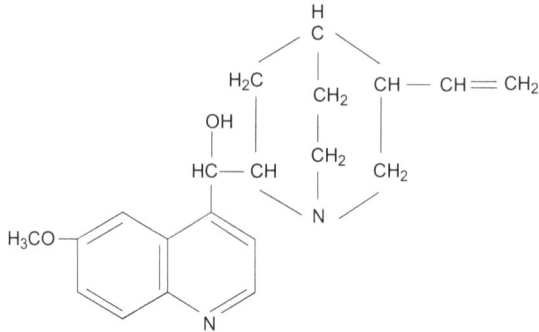

Figure 15. Quinine molecule

<u>Test for the purity of quinine called the Thalleoquine</u>: A few drops of bromine water are added to 5 ml of a 1% solution of quinine in dilute sulfuric acid and the mixture is rendered just alkaline with ammonium hydroxide. An emerald green color is produced. This reaction is given by quinine and quinidine but not by cinchonine and cinchonidine.

<u>Test for the presence of other alkaloids, known as Kerner's Test</u>: This is based on our knowledge that quinine sulfate is more soluble in dilute ammonium hydroxide than the other cinchona alkaloids. Kerner's test is conducted by adding 1 gm of the sample to 30 ml of water contained in a flask that is connected to a reflux condenser. The mixture is boiled until the quinine sulfate has dissolved and then quickly cooled to 15 °C and maintained at that temperature for 30 minutes. The liquid is rapidly filtered off and 5 ml of the Filtrate is transferred to a small beaker on a magnetic stirrer. A 0.5 M ammonium hydroxide solution should be run from a burette while agitating. A white precipitate should form which then re-dissolves yielding a clear solution. Repeat this process and keep record of the amount of ammonium hydroxide used. It should not exceed 6 ml and, if it does, you are probably dealing with a sample with additional alkaloids.

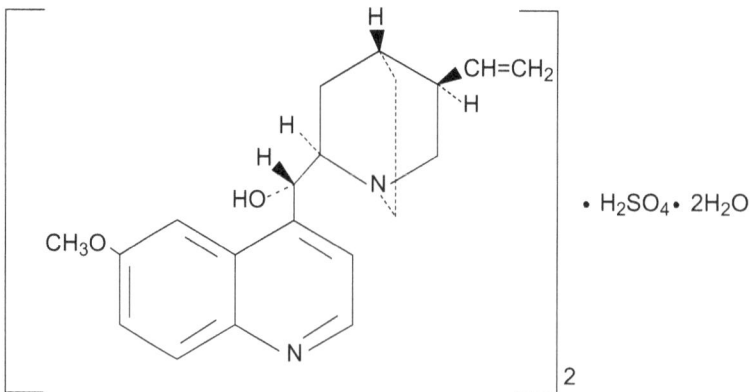

Figure 16. Quinidine Sulfate molecule

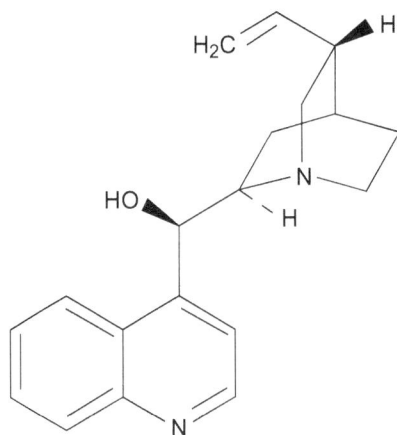

Figure 17. Cinchonine molecule

FEVER BARK

BOTANY – The Fever Bark, known botanically as *Alstonia consyricta*, are ever-green trees which grow to a height of 15 meters. The leaves are glossy, oblong, and petiolate. The tree has a 2-7 cm rusty-brown rugose periderm which is deeply fissured. It is indigenous to Australia, and some alternate species grown in India and the Philippines. The medicinal parts are the bark of the root and tree.

CHEMISTRY – The constituents are mainly indole alkaloids including alstoni-dine, yohimbine, and reserpine.

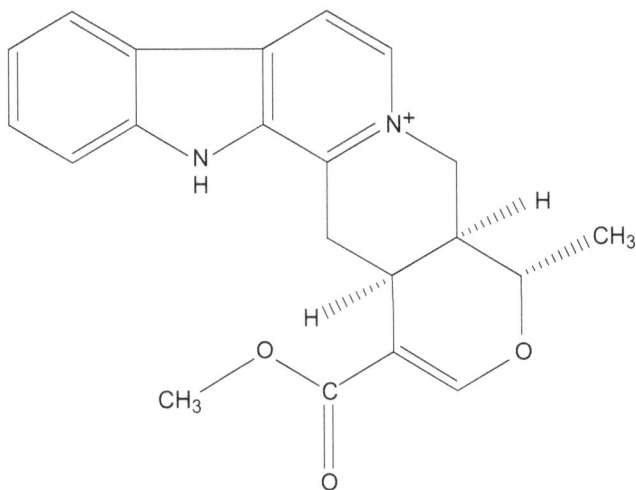

Figure 18. Alstonidine molecule

Alstonidine is an indole alkaloid and shows a specific color reaction. To a 10% solution of the drug add a few drops of sulfuric acid and a 1% solution of ammonium vanadate. The color changes from violet to blue to purple to red. Also with sulfuric acid and potassium dichromate it yields a deep blue color which gradually changes from violet to cherry red to orange and ends in yellow.

WILD INDIGO

BOTANY – The *Baptista tinctoria* is a many-branched plant growing up to 1 meter in height with a woody rootstock and knotty branches. The stem is 1-3 mm thick, round, slightly grooved, and glabrous. The flowers are terminal and axillary in 7-10 cm long lightly flowered racemes. It is indigenous to southern Canada and eastern and northeastern United States. The medicinal part is the root.

CHEMISTRY – Compounds include particular arabino-Galactans, glycoproteins, and quinolisidine alkaloids including cyticine and sparteine.

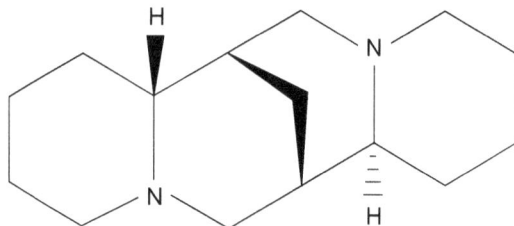

Figure 19. Sparteine molecule

Chemical test for sparteine: Add 25 ml of ether and 2 drops of ammonia TS to 0.1 gm of sparteine sulfate in a test tube. Then add to the mixture an ethereal solution of iodine (l in 50) until the liquid when shaken turns from an orange to a dark reddish-brown color. After a short time the sides and bottom will be coated with dark greenish-brown crystals.

SIMARUBA

BOTANY – *Simaruba amara* is a tree that grows over 18 meters high. The roots are long and spread horizontally. The leaves are 22-27 cm long. The bark that is used commercially is thin and flat. The tree grows in the Caribbean and northern South America.

CHEMISTRY – Compounds consist primarily of quassinoids (breakdown products of tri-terpenes) including simarubin. There are also tannins and volatile oils present.

ADDITIONAL botanicals which may have anti-malarial activity include Azadirachta, Barberry, Croton, and Milk Thistle.

ANTI-MICROBIALS

PHARMACY – Antimicrobials are more a major category or drugs rather than a specific class. A botanical may be regarded as antimicrobial if it attacks both microzoa and multicellular microphytes. So this category includes agents which are bacteriostatic, fungicidal, antiprorozoal, germicidal, and disinfectants which have therapeutic value in numerous areas of the body.

UVA URSI

BOTANY – The *Arctostaphylos uva ursi* plant is a decambent up to l.5 meters long, creeping espalier with elastic, red-brown branches. The leaves are alternate, coriaceous, short petioled spatulate-obovate or wedge shaped, entire margined, and slightly revolate. The plant has spread from the Iberian Peninsula across central Europe northwards to Scandinavia and eastward to Siberia. It can also be found in the Altai Mountains, the Himalayas, and North America. The medicinal part is the leaf.

CHEMISTRY – The constituents of uva ursi are hydroquinone glucosides including arbutin as well as some tannins, flavonoids, and triterpenes.

Identification test for Uva Ursi: Place 0.1 gm, of the drug on a watch glass and covering it with another watch glass. Gently heat the powder. A crystalline sublimation of hydroquinone is formed upon the upper watch glass consisting of long rods and feather-like aggregates which show a brilliant play of colors in polarized light.

Identification test for Uva Ursi purity: Macerate 1 gm of the drug with 10 ml of boiling water. Shake occasionally until cool and then filter. The filtrate will yield a precipitate upon the addition of a few drops of ferrous sulfate TS.

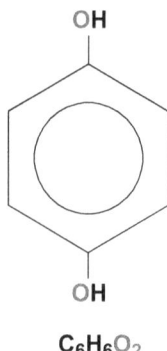

OH

OH

$C_6H_6O_2$

Figure 20. Hydroquinone molecule

TURMERIC

BOTANY – *Curcuma domestica*, known popularly as Turmeric is a perennial erect and leafy plant with very large lily-like leaves up to 1-2 cm long. The-leaf blade is ovate-lanceolate, thin, entire-margined, and narrows to a long sheath-like petiole. Curcumin is indigenous to India but it is also cultivated in other tropical regions of southern Asia. The medicinal parts are the stewed and dried rhizome.

CHEMISTRY – Chief components are and beta tumerone, artumerone, alpha and gamma atlantone, and the curcuminoids including curcumin, demethoxycur-cumin, and bidemethoxycurcumin.

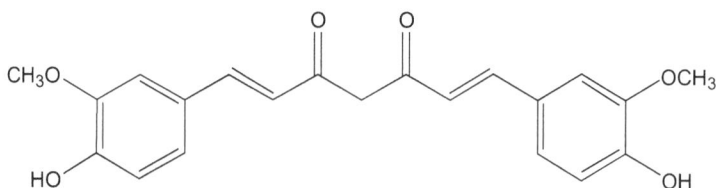

Figure 21. Curcumin molecule

To make turmeric paper for testing: Macerate 20 gm of Turmeric powder with four successive 100 ml portion of cold water. Decant the clear liquid portion each time and discarding it. Dry the residue at a temperature of not over 100 °C. Next macerate with 100 ml of ethanol for several days and filter. Impregnate strips of white filter paper with the resulting solution.

Purity test: This can be conducted by dipping a strip of the turmeric paper into a solution of 1.0 gm of boric acid in 5 ml of water previously mixed with 1 ml dilute hydrochloric acid. After 1 minute, remove the paper and allow to dry. The yellow color should change to brown. Then moisten it with some ammonium hydroxide and the color should change to a greenish-black.

THUJA

BOTANY – *Thuja occidentalis* is a 12-21 meter high tree with short horizontally spread branches and red-brown striped peeling trunk. The leaves are scale-like, cross opposite, imbricate, flattened on the branch side and folded at the margins. The tree originated from eastern North America and is found also in Europe mainly as an ornamental plant. The medicinal parts are the oil extracted from the leaves which is usually collected fresh in the spring.

CHEMISTRY – Constituents include the volatile oil, Thujone plus isothujone and fenchone plus water-soluble immunostimulatin polysaccharides, glyco-proteins, flavonoids, lignans, and tannins.

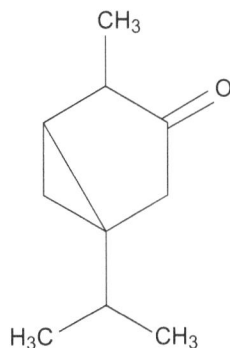

Figure 22. Thujone molecule

The common name for *Thuja occidentalis* is Arbor Vitae. It possesses an oil which can be extracted from the plant leaves by steam distillation.

Extraction method: Place 5 gm of the drug in a mixing flask containing 100 ml of water. Connect this to a steam generator as shown in the figure. The flask with the thuja leaves should then be connected to a condenser and a receiving flask for the volatile oil. Make sure the safety pressure tube is below the surface of the water to prevent excessive pressure and explosion. Begin heating the water in the steam generator until the volatile condensation begins to appear in the receiving flask. Discontinue heating, cool down the apparatus and measure the amount of volatile oil received.

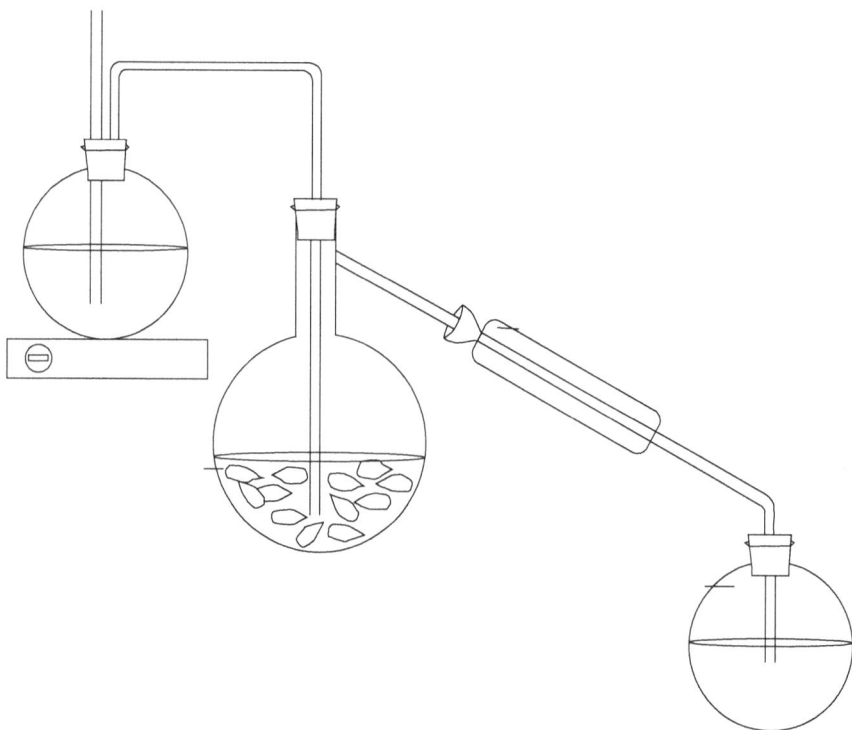

Figure 23. Steam Distillation Apparatus
(**Source:** *Experiments in Pharmaceutical Chemistry*
by Charles Dickson, CRC Press, 2013, p. 110)

To calculate the thujone in your collection of oil, dissolve 5 gm of hydroxylamine hydrochloride in 9 ml of warm water, then add 80 ml of 90% ethanol and 2 ml of bromophenol blue TS. Neutralize this mixture with 0.5 M potassium hydroxide (alcoholic) and add sufficient ethanol to make 100 ml. Then weigh 1 gm of thuja oil and 20 ml of hydroxylamine hydrochloride. Shake the mixture and titrate it with 0.5 M potassium hydroxide in ethanol. Continue titration for 4 hours at intervals of 30 minutes. Each ml of the 0.5 M potassium hydroxide used is equivalent to 0.0761 gm of the ketone known as thujone.

CLOVES

BOTANY – Clove, known botanically as *Syzygium aromaticum*, is a tree that grows up to 20 meters high, is pyramid-shaped, and evergreen. The diameter of the trunk can be as much as 40 cm. The branches are almost round and the leaves are 9-12 cm long and 3.5 cm wide, coriaceous, elliptical to lanceolate, short, obtusely tipped and narrowing in a cuneate form to the petiolate. Clove is indigenous to the Molucca Islands and is also cultivated in Tanzania, Madagascar, Brazil, and other tropical regions.

CHEMISTRY – The volatile oil is eugenol along with eugenyl acetate and betacarophyllene.

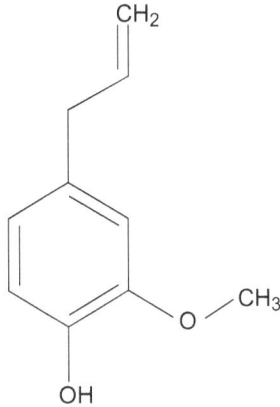

Figure 24. Eugenol molecule

Test to show the hydrocarbon part of eugenol: Dissolve 1 ml of eugenol in 20 ml of a 0.5 M sodium hydroxide solution in a 50 ml stoppered tube. Add 18 ml of water and the solution becomes clear.

Test to illustrate the phenol part: Shake 1 ml of eugenol with 20 ml or water. Then add a drop of ferric chloride TS and watch for the transient grayish green color.

CLEMATIS

BOTANY – The *Clematis recta* plant grows from 50-125 cm high. The stem is non-climbing, erect, leafy, and glabrous. The leaves are pinnatifid and the white flowers are in many-blossomed terminal cymes. The medicinal part is the fresh flowering part.

CHEMISTRY – Clematis contains protoanemonine-forming agents probably the glycoside, ranunculin. An infusion used as a paste for a dermal poultice can be made by placing 10 gm of the dried clematis powder in 100 ml of water that is boiling. Allow the mixture to set for 30 minutes and strain off the resulting liquid.

ADDIITIONAL botanicals which possess various forms of microbial action include Yarrow, Mountain Laurel, Scarlet Pimpernel Butternut, Bistort, and Aloe.

APHRODISIACS

PHARMACY – Aphrodisiacs are substances which possess the power to stimulate the genital organs. Such type of action varies according to the drug but such stimulation may include a mildly positive estrogen reaction on the genital system, orgasm stimulation, or the increase of potency.

YOHIMBE

BOTANY – The *Pausinystalia yohimbe* is an evergreen that grows up to 30 meters high in the jungles of West Africa, Cameroon, Congo, and Gabon. The bark is gray-brown fissured and split. It is often spotted. The inner fracture is reddish-brown and grooved. The leaves are oblong or elliptical. The medicinal part is the bark.

CHEMISTRY – Constituents are mainly indole alkaloids including yohimbine (quebrachine) and its numerous stereoisomers.

N

Figure 25. Indole Alkaloid Nucleus

Yohimbine (quebrachine) is an indole alkaloid that exhibits the characteristics of that class.

Test: When sulfuric acid and a 1% solution of ammonium vanadate are added to a 5% solution of yohimbine, the color changes on standing from violet to blue to purple, and finally to red.

If 1 ml of sulfuric acid and potassium dichromate are added to a solution of yohimbine, a deep blue color appears that gradually changes to violet.

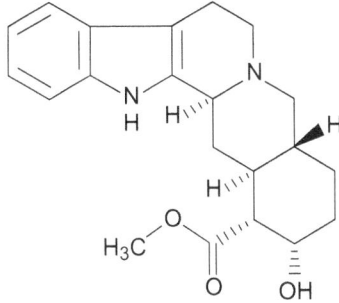

Figure 26. Yohimbine molecule

DAMIANA

BOTANY – The *Turnera diffusa* plant is a small shrub which grows up to 60 cm high. The leaves are 1-2.5 cm long. They are smooth and pale green on the upper surface glabrous with a few scattered hairs on the ribs between. The leaves are ovate-lanceolate short petioled and have two glands at the base. The medicinal parts are the leaves harvested during the flowering season. The plant is found mainly in the region of the Gulf of Mexico, the Caribbean, and southern parts of Africa.

CHEMISTRY – The chief components are alpha-pinene, beta-pinene, 1,8 cineole, and para-cymene. There are also tannins (4%), resins (7%), plus hydroquinone glycosides (arbutin) and cyanogenic glycosides (tetraphylline B, barterin).

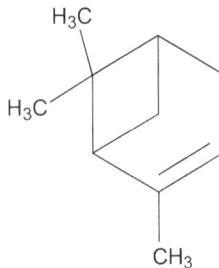

Figure 27. Alpha-Pinene molecule

NETTLE

BOTANY – With the scientific name of *Urtica dioica*, the plant grows from 60-150 cm high. The leaves are opposite, oblong-cordate and roughly serrate. The whole plant is covered in stinging hair, hence the popular name of Stinging

Nettle. The medicinal parts are the fresh and dried flowering plant and the roots. The plant is most common in temperate regions of the world.

CHEMISTRY - Compounds present in Nettle include Sterols such as beta sitosterol, stigmasterol and compesterol plus other sterols. In addition there are also lactins, water-soluble polysaccharides, (glucans, glucogalacteronans, and arabino galactans) plus hydroxycourmarins, lignans, and ceramides.

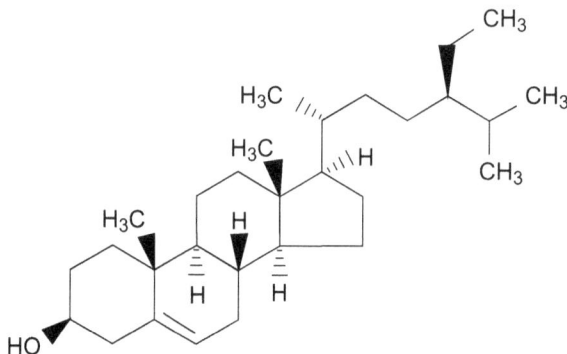

Figure 28. Beta-Sitosterol molecule

CAYENNE

BOTANY – Cayenne, known botanically as *capsicum annum*, is an annual (in the tropics, perennial) plant that grows from 20 to 100 cm high with an erect stem which is somewhat woody and angular. The leaves are usually solitary, long-petioled, oval, lanceolate to ovate obtuse-accuminate wedge-shaped at the base entire-margined or slightly curved and glabrous. The plant is indigenous to Mexico and South America and cultivated in many other areas. The medicinal part is the fruit (red pepper)

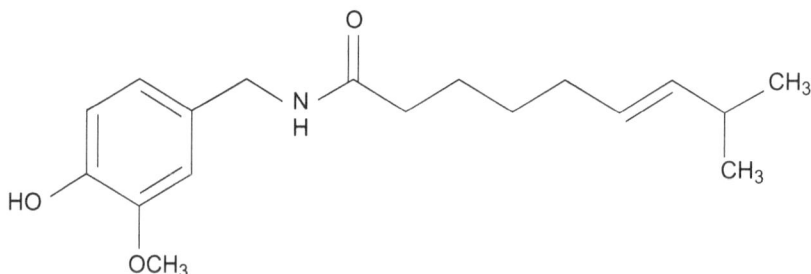

Figure 29. Capsaicin molecule

CHEMISTRY – Primarily composed of capsaicinoids (amides of the vanillyl amine with C8 to Cl3 fatty acids). The chief component is capsaicin but there are also carotinoids, flavonoids, and steroid saponins.

Purity test for capsaicin: Take 1 gm of capsaicin powder with 50 ml of ethanol, place in a stoppered flask, and allow to macerate for 24 hours. Dilute 0.1 ml of the clear supernatant liquid in 140 ml of distilled water containing a 10% solution of sucrose. Five ml of this dilution swallowed slowly should produce a distinct sensation of the pungency of capsicum.

ADDITIONAL botanicals reputed to have aphrodisiac qualities include Gingko biloba, Horny Goat Weed, Cinnamon, Epimedium Leaf, and Stinging Nettle.

ASTRINGENTS

PHARMACY – Astringents are locally applied protein precipitants which have such a low cell penetrability that the actions are essentially limited to the cell surface and interstitial spaces. When the tannins precipitate protein from solution and are able to combine with them, rendering them resistant to proteolytic enzymes.

WITCH HAZEL

BOTANY – Witch Hazel, or *Hamamelis virginiana*, is a treelike deciduous bush that grows 2-10 meters high with a trunk diameter up to 40 cm. The bark is thin brown on the outside and reddish on the inside. The leaves are alternate with stipules and the margin is roughly crenate, bluntly indented to irregular sweeping. The medicinal parts are the bark and the leaves. The tree is from the deciduous forest of Atlantic regions of North America. It also grows in Europe and is cultivated in subtropical countries.

CHEMISTRY – The compounds are mainly tannins including hamamelitannin, monogalloylhammeloses and oligomeric procyanidins.

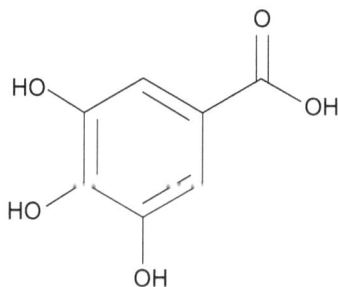

Figure 30. Gallic Acid molecule

Tests: In test tube # l add a drop of ferric chloride TS to 5 ml of witch hazel. Note the brownish-yellow color which develops.

In test tube #2, add 2 drops of potassium ferricyanide dissolved in ammonium hydroxide to 5 ml of witch hazel and note the deep red color.

To a 20% gelatin solution in test tube #3, add 1 ml of witch hazel and note the protein precipitation.

To test tube #4, add 1 ml of potassium dichromate TS to 5 ml of witch hazel and note the precipitate.

 In test tube # 5 place an ml of lead acetate TS in 5 ml of witch hazel and check the precipitate.

CITRONELLA

BOTANY – The *Cynbopogon citatus* is a perennial plant with up to 2 meters glabrous stalk. The leaf blade is linear acuminate, up to 90 cm long, 5 mm width and smooth on both sides. The leaf blade is up to 1 meter long and 1.5 meters wide and usually light green. The inflorescence is very large and consists of 1 meter long spikes with numerous raceme up to 20 mm long and arranged in zigzag order. The medicinal parts are the dried leaves. Citronella was originally indigenous to the tropics and subtropics of the old world. Today it is cultivated in Central and South America and in Queensland, Australia.

CHEMISTRY – Chief components are volatile oils including citral and myrcene if the origin is citrates or citronellal, geraniol, citronellol, geranylacetate, and citronellyl acetate.

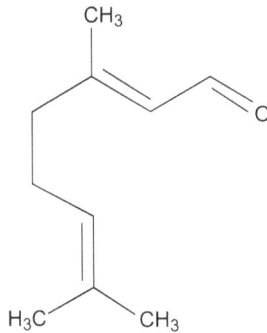

Figure 31. Geraniol molecule

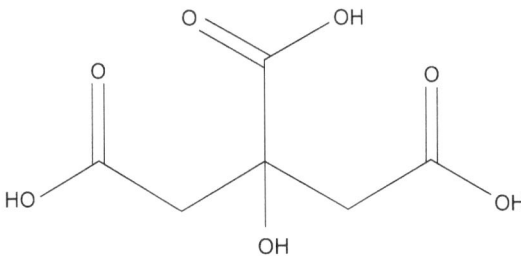

Figure 32. Citric Acid molecule

Test: To conduct an assay on citric acid which is present in Citronella products, place 2 gm of citric acid with Sulfuric acid in a dessicator for 5 hours. Then place the powder in 100 ml mix of glycerin and water that has been previously neutralized using phenolphthalein TS. Add phenolphthalein TS and titrate with 1.0 M sodium hydroxide. Now discharge the pink color by the addition of 50 ml of glycerin and continue the titration with sodium hydroxide until the pink reappears. Each ml of the 1.0 M sodium hydroxide used is equivalent to 64.04 mg of citric acid.

OAK BARK

BOTANY – The *Quercus robur* tree grows to about 50 meters high with a broad, irregular, heavily branched crown and a trunk which divides into gnarled, strong bent branches. The bark is deeply fissured, thick, and gray-brown. The leaves are short petioled almost sessile, oblong-obovate, almost lobed, and usually cordate at the base. The medicinal parts are the dried bark of the trunk and branches, the dried leaves of various oak varieties and the seeds without the skin. The tree is widespread in Europe, Asia Minor, and the Caucasus region.

CHEMISTRY – Components are about 16% various kinds of tannins including, among others castalagin, pedunculagin, vesvalagin, flavano-ellagic tannins (acutissimine A and B, eugenigrandin, guajavacine B, and stenophyllanin). There are also catechu tannins including oligomeric proanthocyanidins plus monomeric and dimeric catechins and leucocyanidins.

Test: Place 1 gm of oak bark powder in a flask of 20 ml of water. Add l ml of ferric chloride TS. A characteristic bluish-black color appears.

Test: Place 2 ml of oak bark in a water solution in a flask and add 1 ml of freshly-prepared gelatin solution. The tannins will produce a precipitate in the gelatin solution.

Test: Place a few drops of 1% ferric acetate TS in a beaker with 10 ml of oak bark water solution. An orange-brown color should appear.

Test: Place a few drops of 1% sodium carbonate TS in a beaker holding 20 ml of oak bark water. This should result in a yellowish-brown precipitate.

ADDITIONAL botanicals with astringent properties include nutgall, catechu, tannic acid, chestnut, sumac acacia, and kino.

CARDIOVASCULAR

PHARMACY – Cardiovascular covers, as the term implies, heart and circulatory system. Drugs create a great variety of reactions in this system and are therefore used in a variety of ways to achieve different therapeutic purposes. Of course, there are many subclassifications under cardiovascular including drugs that may be hypertensive or anti-hypertensive (hypotensive), ganglionic blocking agents, cholinergic agents, vasodilators and vasoconstrictors among others.

DIGITALIS

BOTANY – *Digitalls purpurea* is a biennial with branches and a tap root. In the first year it develops a leaf rosette. In the second year it produces a two-meter high erect, unbranched, gray, tomentose stem. The leaves are alternate, ovate, tapering upwards, and petiolate. The plant is indigenous to Europe but is also cultivated in Asia and North America.

CHEMISTRY – Constituents include glycosides (aglycone, digoxigenin, which are primary glycosides) plus digitoxin, a secondary glycoside and various saponins and anthraquinones.

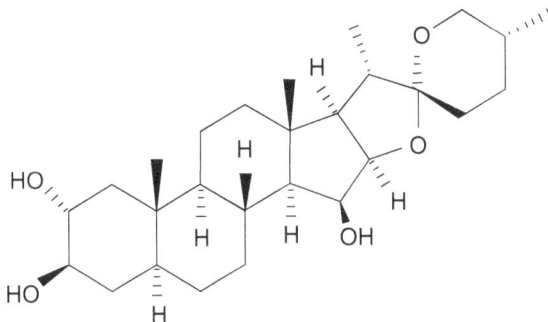

Figure 33. Digitogenin molecule

Test: Place 1 gm of digitalis leaf powder in 50 ml of concentrated hydrochloric acid. Note the identifying reddish-brown color which appears.

Keller's Reaction: Dissolve 1 gm of digitalis leaves in 10 ml of glacial acetic acid in a large test tube. Add a few drops of ferric chloride TS and then gently add some sulfuric acid. The characteristic color reaction includes a brown ring which appears at the juncture of the fluids.

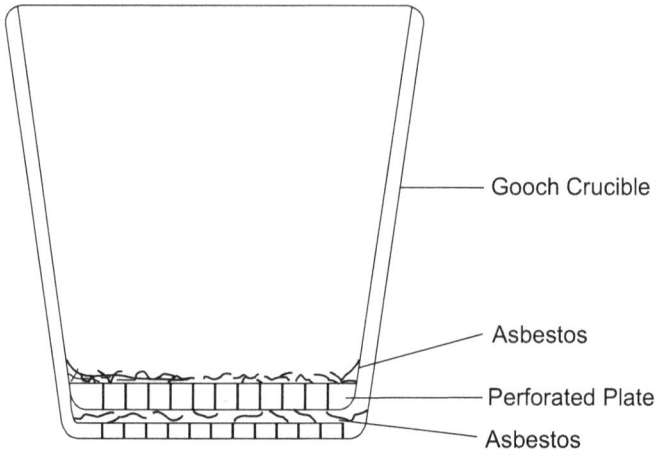

Figure 34. Gooch crucibles
(**Source:** Wikipedia)

Ash Content: Determine the ash content of digitalis leaves by weighing 4 gm of the leaves and placing it in a tared Gooch crucible. Then incinerate at a low temperature not to exceed very dull, redness. Cool the crucible and ash in a dessicator and, when cool, weigh the ash. The ash content of the digitalis leaves should calculate to be about 92%.

Acid-Insoluble Ash Content: Determine the acid insoluble ash content of the leaves by placing the ash powder from the previous procedure with 25 ml of dilute hydrochloric acid for a period of about 5 minutes. Collect the insoluble material on a tared Gooch crucible. Wash it over with hot water, ignite, and weigh. The percentage of acid-insoluble ash in the digitalis leaves should calculate out to be about 8%.

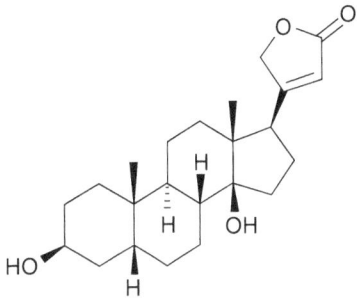

Digitoxigenin

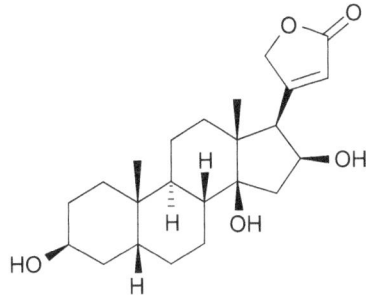

Gitoxigenin

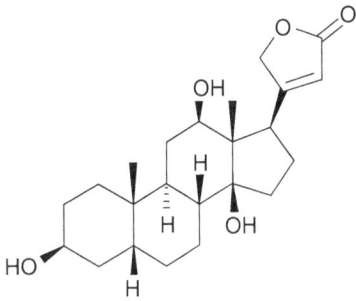

Digoxigenin

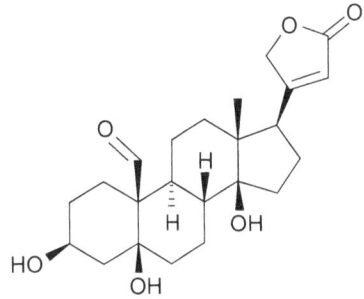

Strophanthidin

Figure 35. The Cardiac Glycosides of Digitalis

APOCYNUM

BOTANY – *Apocynum cannabinum* also called Canadian Hemp is a perennial growing up to two meters high. It has an erect stem which branches at the top. The short petioled leaves are 5-11 cm long, yellowish-green, and oblong or oblong-ovoid. The tips of the leaves are initially rounded and then terminate abruptly in thorny tips. The medicinal parts are the root and the juice obtained from the fresh plant. It grows mostly in the United States and Canada but is also cultivated in Russia.

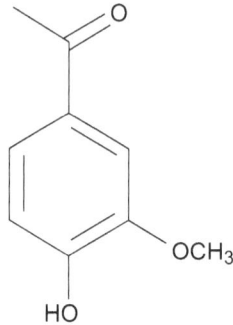

Figure 36. Apocynin molecule

CHEMISTRY – The chief compounds are steroid glycosides particularly, cymarin, K-strophantoside, apoannoside and cyanoside.

Acid-insoluble ash: The ash is obtained by burning the apacynum in a crucible and coalingo. Then place 2 gm, of the ash in a flask containing 25 ml of hydrochloric acid. Collect the insoluble material by use of filtration and place the filtrate in a tared Gooch crucible. After heating and cooling with hot water, ignite, and weigh. The acid-insoluble ash in apocynum should calculate to be about 5%.

STROPHANTHUS

BOTANY – Strophanthus is a plant climbing lianes which is occasionally erect shrub or tree. The leaves are opposite, ovate to elliptical, short-petioled, simple entite margined and usually coriaceous. The plant is native to India and tropical parts of the world.

CHEMISTRY – Cardiovascular steroids constitute the plant, particularly strophanthix, strophanthidin, and acolongifloriside plus saponins and fatty oils.

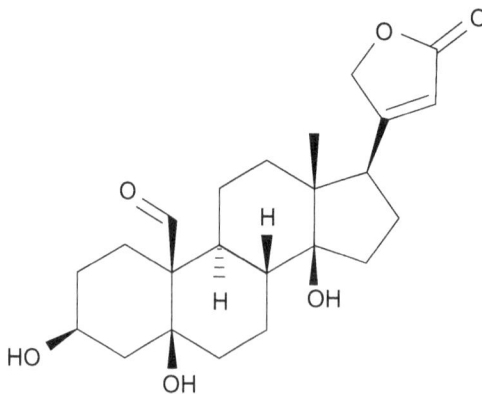

Figure 37. Strophanthidin molecule

Test: To an aqueous solution of strophanthin add 2 ml of sulfuric acid and a few drops of ferric chloride TS. A red precipitate occurs which becomes dark green in an hour.

Test: Mix an aqueous solution of strophanthin with 10 ml of dilute hydrochloric acid and heat the mixture to 70 °C. The strophanthin is hydrolyzed into strophanthidin which precipitates.

SQUILL

BOTANY – *Scilla*, sometimes called White Squill or Red Squill, depending upon species and area. The color refers to the outer membranous skin of the bulbs from which the drug is extracted. Powdered squill is yellowish-white to very pale brown. It is very hygroscopic, caking in a moist atmosphere. The upper and lower epidermal layers are elongated thin-walled cells with a few elliptical stomata and a mesophyll of polygonal thin-alled parenchyma cells. Squill, depending upon variety is indigenous to the Mediterranean region but is cultivated in other parts of the world.

CHEMISTRY – The active ingredients of Squill are glucosides particularly scillaren A and scillaren B.

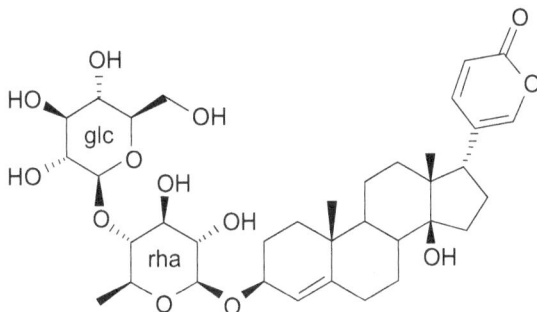

Figure 38. Scillaren A molecule

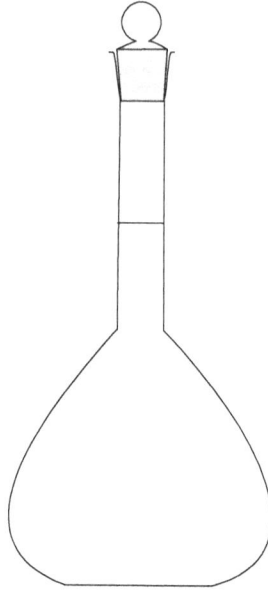

Figure 39. Glass Stoppered Volumetric Flask
(**Source:** H. Padleckas drawing on *Wikipedia*)

<u>Tincture of Squill</u>: Place 10 gm of Squill powder in a 200 ml volumetric flask with a menstruum of ethanol-water 3:1. Percolate slowly and allow to macerate over a period of 24 hours. After this period of time, filter and bottle the tincture which should be about 64-67% ethanol by volume.

BARBERRY

BOTANY – Barberry, known botanically as *Berberis*, is a deciduous heavily-branched thorny bush which grows up to 2 meters high. The thorny branches are angular, deep-grooved, initially brownish-yellow and later more whitish gray. The leaves are in bushels and obovate to elliptoid, 2-4 cm long and narrowed into a 1 cm long petiole. The leaves are dark green and reticular. The habitat ranges from Europe to northern Africa as well as parts of America and central Asia. The medicinal parts are the fruit and root bark.

CHEMISTRY – Constituents include isoquinoline alkaloids, particularly berberine plus anthocyans and chlorogenic, malic, and acetic acids.

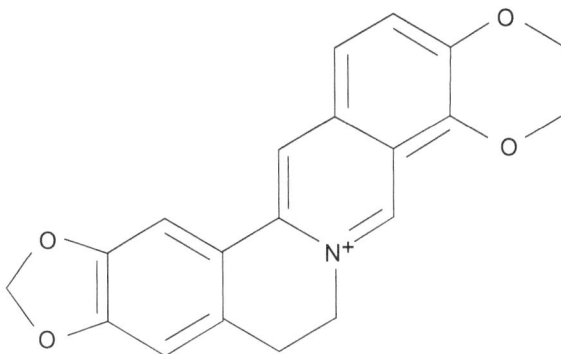

Figure 40. Berberine molecule

<u>Isolating berberine from the botanical source</u>: Moisten 10 gm of barberry root or leaf with sufficient alkaline mixture of water, ethanol, and ethyl ether (1:1:8) to become damp and allow the mixture to stand in a covered container for 45 minutes. The drug is then transferred to an extraction thimble and the drug is extracted with ether in a Soxhlet apparatus for at least six hours. Following that step, the etheral extract is placed on a filter paper and time allowed for the solvents to evaporate. Add 50 ml of water to the residue in a 100 ml beaker and boil to dissolve the berberine. Filter the solution while it is still hot through cotton placed in a funnel. Next add 5 ml of 5% hydrochloric acid until the solution is acidic.

Berberine hydrochloride will settle out as dark crystals. Filter off the crystals, dry, and weigh.

Compare the weight of the filtrate with the original 10 gm with which you begun the experiment. Berberine and a few closely-related alkaloids can occupy as much as 80% of the weight of the root bark.

SAFFLOWER

BOTANY – Safflower is known botanically as *Canthamus tinctoria* which is a plant that grows up to 90 cm high with a fusiform root. The stem is erect, simple, or branched. The leaves are long, fairly soft or glabrous with a thorny-serrate margin and tip. The plant oil extracted from its embryo. The plant is indigenous to Iran and northwest India but is also cultivated in the United States.

CHEMISTRY – The constituents can range as high as 55-85% in fatty acids, chiefly linolenic and linoleic acids, both of which are polyunsaturated.

<u>Test</u>: The most rapid test for determining unsaturation is to add a few drops of iodine solution to about 10 ml of safflower oil. Within a few minutes the iodine loses its color due to addition across the double bonds, thus going from an alkene to an alkane.

Formula: $CH_3-(CH_2)_4-CH=CH-CH_2-CH=CH-(CH_2)_7-COOH$

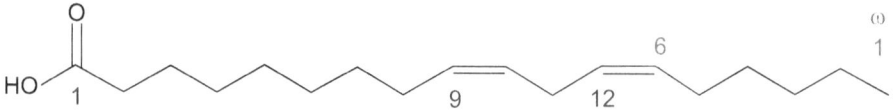

Figure 41. Linoleic Acid molecule

VERATRUM

BOTANY – The botanical, *Veratrum albin*, known as White Hellebore is a plant which grows 60-120 cm high. The rhizome is short, cylindrical, stunted and has numerous long flesh root fibers. The stem is almost completely surrounded by a tight sheath of basal leaves. The leaves are whorled, broad, elliptical to linear lanceolate and heavily ribbed. The plant is found in Europe from Lapland in the north to Italy in the south. The medicinal parts are the rhizome and the roots.

CHEMISTRY – Most of the components are steroid alkaloids including protover-ine and protoveratrine. Test for acid insoluble ash by placing the weighed ash in a solution of dilute hydrochloric acid for 5 minutes. Collect the insoluble matter on a tared Gooch crucible, wash with hot water, ignite, and weigh. The acid-insoluble should be no more than 4%.

ADDITIONAL botanicals with cardiovascular actions include Adonis, Shepherd's Purse, Lily of the Valley, Bishop's Weed, and Hawthorn.

CARMINATIVES

PHARMACY – Carminatives are substances which relieve gaseous distention in stomach or intestines and may also act to increase gastric and salivary secretion, thus to improve appetite. Carminatives also include a number of flavoring agents whose active ingredient is a volatile oil. Carminatives are frequently incorporated into medicines into medicines by diluting to volume with the appropriate water the drug as a substitute for ordinary distilled water.

GINGER

BOTANY – Ginger, known botanically as *Zingiber officinale* is a creeping perennial on a thick tuberous rhizome which spreads underground. In the first year a green erect, reed-like stem about 60 cm high grows from this rhizome. The plant has narrow, lanceolate to linear lanceolate leaves 15-30 cm long. The flower scape grows directly from the root and terminates in a long curved spike. A white or yellow flower grows from each spike.

CHEMISTRY – *Zingiber officinale* contains arylalkanes and volatile oils such as zingiberene and arcurcumene. There are also gingerdiols and diarylheptanoids.

Test: To determine the presence of sulfites in ginger 500 ml of water and 20 ml of hydrochloric acid are placed in a large flask and 10 ml of hydrogen peroxide, previously neutralized. Bromophenol blue indicator is added to each absorption tube. The apparatus is connected up and a stream of carbon dioxide previously bubbled through sodium carbonate solution is passed in while the dilute acid is heated to boiling. After boiling has continued long enough to remove air, the flask is cooled by running a stream of cold water over the outside. The stopper is removed, 100 gm of powdered ginger is quickly introduced and the stopper replaced. The mixture is gently boiled for at least 45 minutes. The absorption tubes are then disconnected and their contents titrated with 0.1 M sodium hydroxide using bromophenol blue as the indicator. Each ml of the 0.1 M sodium hydroxide required in titration is equivalent to 0.0032 gm of sulfur dioxide. In the absence of sulfites in the ginger, no acidity will have developed.

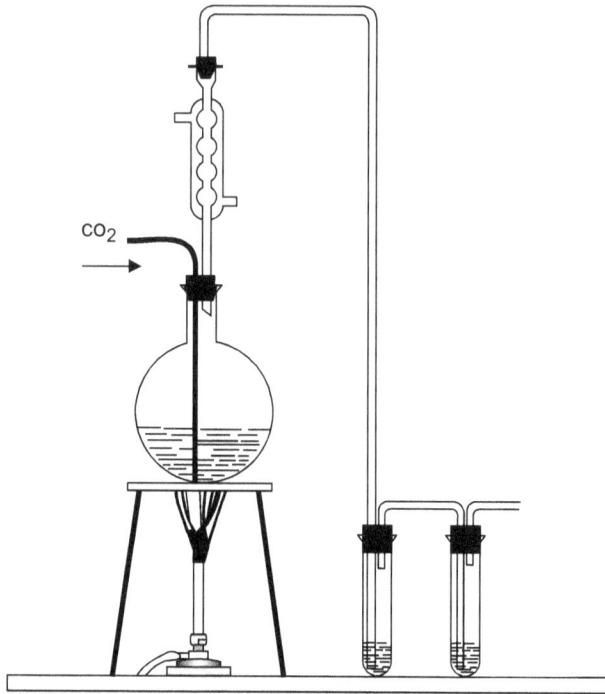

Figure 42. Distillation Apparatus

<u>Test - water soluble ginger</u>: Place 4 gm of ground Ginger in a 200 ml volumetric flask, fill to the mark with water and agitate, using a magnetic stirrer at 30-minute intervals during an 8-hour period. Allow the mixture to stand for an additional 16 hours and then filter. Evaporate 50 ml of the filtrate, representing 1 gm of the drug on a water bath and dry the residue at 105 °C for 2 hours. The weight of the extract should not be less than 120 mg.

<u>Test for ether-soluble extract</u>: Place 20 gm of powdered ginger in an extraction thimble of a

Soxhlet extractor and extract with ethyl-ether for six hours. Evaporate the ether extract on a steam bath until the odor of ether is no longer perceptible. Next, dry the residue in a dessicator over sulfuric acid for 18 hours. The weight of the extract so obtained should not be less than 900 mg.

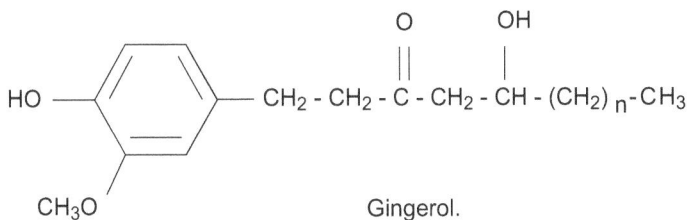

Figure 43. Gingerol molecule

CARDAMON

BOTANY – Cardamon is called *Elettaria cardamomum* and is a perennial with thick tuberous rhizome and numerous long roots. There are up to 30 erect, glabrous green stems 2-3 meters high. The leaves are in two rows with a leaf membrane at the end of the soft-haired sheath.

The medicinal parts are the oil extracted from the seeds and the fruit.

CHEMISTRY – The most important constituents are the volatile oils including the ester, terpinyl acetate and the corresponding alcohol, terpine

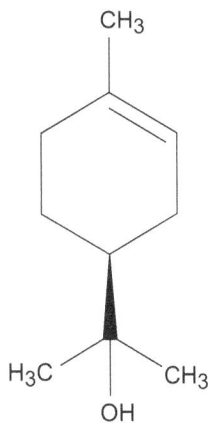

Figure 44. Terpineol molecule

The most important preparation is the compound tincture which his prepared by mixing together equal weights of powdered cardamom seed, and caraway fruit with cinnamon.

CINNAMON

BOTANY – *Cinnarnomum aromaticum* is an evergreen tree which grows up to 7 meters tall with aromatic bark and angular branches. The bark is brown in quilled pieces. The 7.5-10 cm long leaves are lanceolate and are on 6-8 cm long petioles. The flowers are small and on short, slender, silky pedicles. The tree is indigenous to southern China, Vietnam, and Myanmar. The medicinal parts are the flowers collected and dried after they have finished blooming and the partly peeled bark of the branches.

CHEMISTRY – The chief constituents are volatile oils particularly cinnamaldehyde, cinnamyl acetate, coumarin, cinnamyl alcohol, and o-methoxy cinnamylaldehyde.

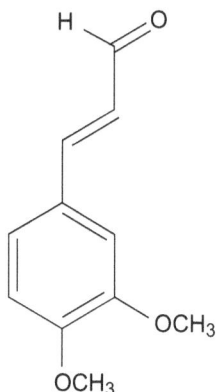

Figure 45. Cinnamylaldehyde molecule

Test: Volatile oil in cinnamon can be recovered by placing 10 gm of powdered cinnamon in the boiling flask as in the diagram and fill the flask half full of water. Attach the condenser and the proper separator. Use glass beads to prevent bumping. Boil for 2 hours until the oil is separated and no longer collects in the graduated tube. Calculate the percentage of oil on the basis of the amount of drug used.

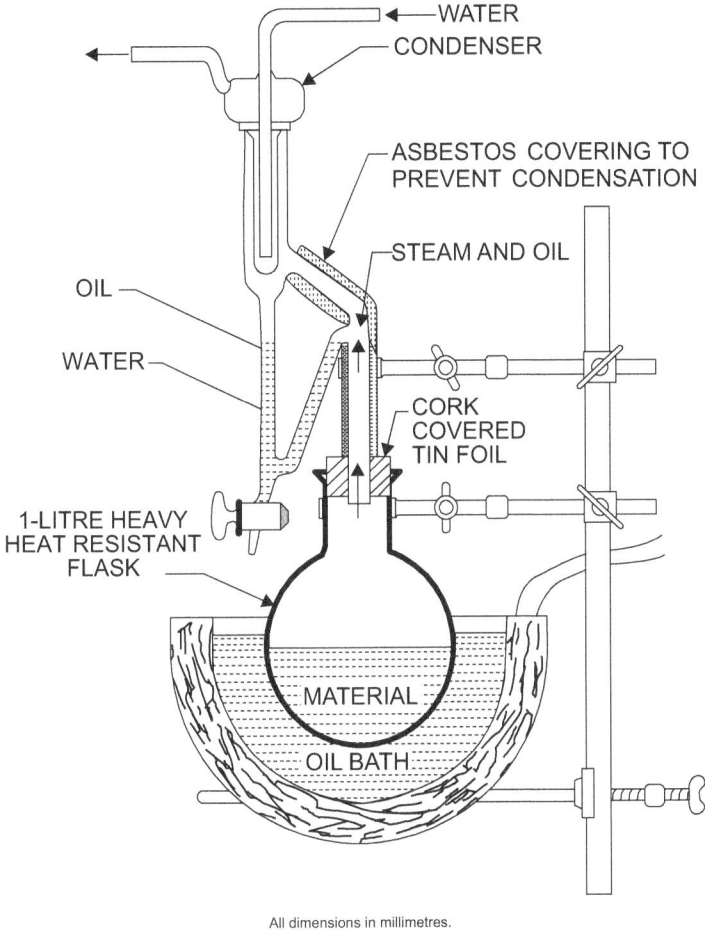

WATER
CONDENSER
ASBESTOS COVERING TO
PREVENT CONDENSATION
STEAM AND OIL
OIL
WATER
CORK
COVERED
TIN FOIL
1-LITRE HEAVY
HEAT RESISTANT
FLASK
MATERIAL
OIL BATH

All dimensions in millimetres.

Figure 46. Volatile Oil Collection Apparatus
(**Source:** *Indian Standard METHODS OF TEST FOR SPICES AND CONDIMENTS*,
https://law.resource.org/pub/in/bis/S06/is.1797.1985.html)

PEPPERMINT

BOTANY – The popular Peppermint, known botanically as *Mentha piperita*, is a perennial plant, 50-90 cm high. The usually branched stems are glabrous but sometimes gray, tomentose, and often oblong-ovate and serrate. The plant is common in both Europe and the United States. Medicinal parts are the oils extracted from all parts of the plant.

CHEMISTRY – The chief volatile oil is menthol and related oils plus some flavenoids.

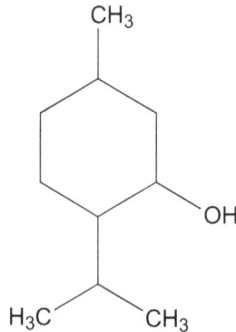

Figure 47. Menthol molecule

Test: To assay for volatile oils place 10 ml of the oil in a 250 ml flask and add 5 ml of neutralized ethanol. Add dropwise into the mixture 0.1 M sodium hydroxide until a pink color appears. Add 25 ml of 0.5 M potassium hydroxide (alcoholic), connect the flask to a reflux condenser, and heat on a boiling water bath for an hour. The final volume of the oils should be about 45% of the original volume of the oil.

SPEARMINT

BOTANY – Called *Mentha spicata*, the plant is 30-60 cm high during the flowering season. The spike-like inflorescence consists of false whorls in the axils of the bracts. The 5-tipped calyx is campanulate, glabrous, or pubescent surrounded by a 5-tipped pale lilac, pink or white corolla. The plant is indigenous to the Mediterranean region but naturalized in Europe and United States. Medicinal parts are the flowers and dried leaves.

CHEMISTRY – Chief components are the volatile oils, primarily Carvone but also limonene.

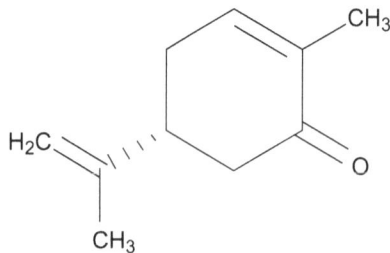

Figure 48. Carvone molecule

ADDITIONAL botanicals with carminative effects include coriander, dill, chamomile, and anise.

DEMULCENTS

PHARMACY – Demulcents are protective agents which are employed to alleviate irritations of mucous membranes and abraded tissue. Some are used internally and others only externally. The local action of chemical, mechanical, or bacterial irritants is thereby diminished. Demulcents are frequently medicated. In such instance the demulcent may be an adjuvant, a corrective or a pharmaceutical necessity. A variety of chemical substances possess demulcent properties including mucilages, gums, dextrins, starches, some sugars, and polymeric polyhydric glycols.

GUM ACACIA

BOTANY – *Acacia senegal*, also called Gum Arabic, is a tree which grows up to 6 meters high with a 12-25 cm thick trunk which is usually slightly leaning with knotty branches and a thin crown. The leaves are double abruptly pinnate and the leaflets are in 10-15 pair narrow gray-green, up to 5 cm long and very short petioled. The medicinal parts are the trunk and branches which supply latex. The tree grows throughout the continent of Africa.

CHEMISTRY – The constituents are mainly polysaccharides and a few glycoproteins.

Test: Add 1 ml of ferric chloride TS to 10 ml of a 2% solution of Acacia. Note that no blackish color appears as does in tannin-bearing gums.

Test: Boil 10 ml of a 2~.Acacia solution and add some drops of iodine TS. Note that no reddish or bluish color appears as does with starch or dextrins.

Test for Acid-Soluble Ash: Dissolve 10 gm of powdered Acacia in about 159 ml of distilled water held in a 250 ml Erlenmeyer flask. Then add 20 ml of dilute hydrochloric acid and boil the mixture for 15 minutes. Filter the hot mixture with suction into a Gooch crucible. Wash thoroughly with hot distilled water and dry at 100 °C. Weigh the obtained residue which represents acid-insoluble ash. It should not weigh more than 0.1 gm if the sample was pure Acacia.

AGAR

BOTANY – *Gelidium amansii* is a perennial seaweed which grows to 1 meter long. The thallus sprouts from a permanent base every year and is heavily branched. It can be cylindrical or flattened, pinnately subdivided and tough. The medicinal part is the seaweed extract known as Agar. The plant is indigenous to the Pacific coasts of Japan and China Sri Lanka, and the South African coast.

CHEMISTRY – The compounds are mainly D-galactose and 3,6 anhydro-L-galactose.

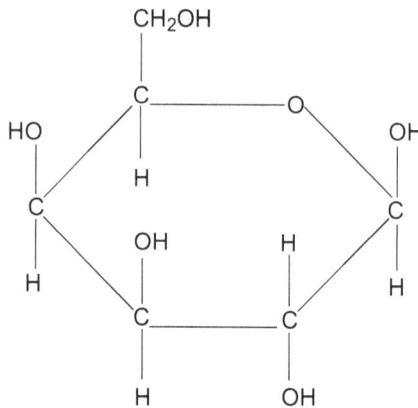

Figure 49. Galactose molecule

Test: Boil 5 ml of Agar in a small beaker. Cool and add some drops of iodine TS. Some fragments of the agar will turn bluish-black while some will turn reddish-violet.

Test: Dissolve 1 gm of agar in 100 ml of water. Boil 5 minutes and allow to cool. Add 5 ml of picric acid TS, and note that no turbidity occurs which makes it different from gelatin.

GUMS IN COMPOUNDING PHARMACY

Comparing solubilities of Agar, Acacia, Karaya, and Tragacanth

The varying stabilities of gums in different solvents are an important consideration in pharmaceutical compounding. Set up an experiment using twelve 125 ml Erlenmeyer flasks labelled as illustrated.

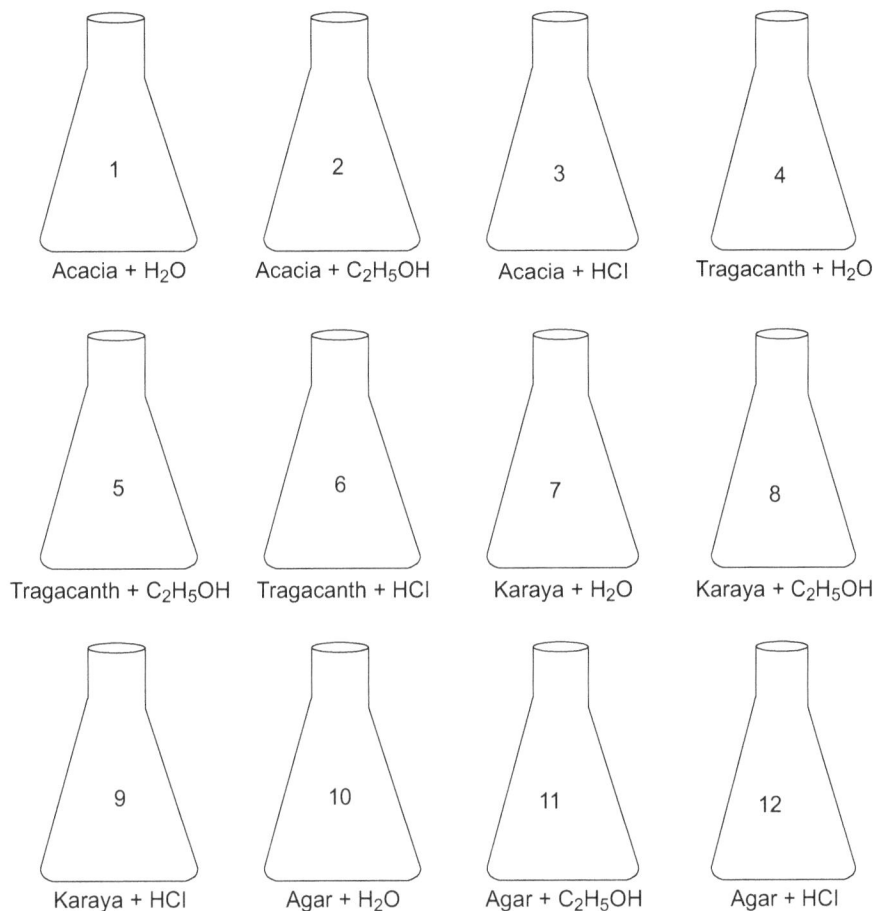

Figure 50. Erlenmeyer Flasks for Gums Experiment
(**Source:** *Experiments in Pharmaceutical Chemistry* by Charles Dickson, CRC press, 2013, p. 102)

Experiment procedure: Introduce 2 gm of each of the four gums into flasks that contain water in the first of each of the four series, then ethanol in the second flask of each of the four groups, and finally, hydrochloric acid in the third flask of each series. The next step is to set up a filtration funnel rack using the previously weighed pieces of cheesecloth. Record all the original weights of the cheesecloths in the data. After filtration is complete place the filters with residues in a drying oven at 103 °C for one hour. After this remove from the oven and allow an hour for them to cool. Then weigh each filter and residue.

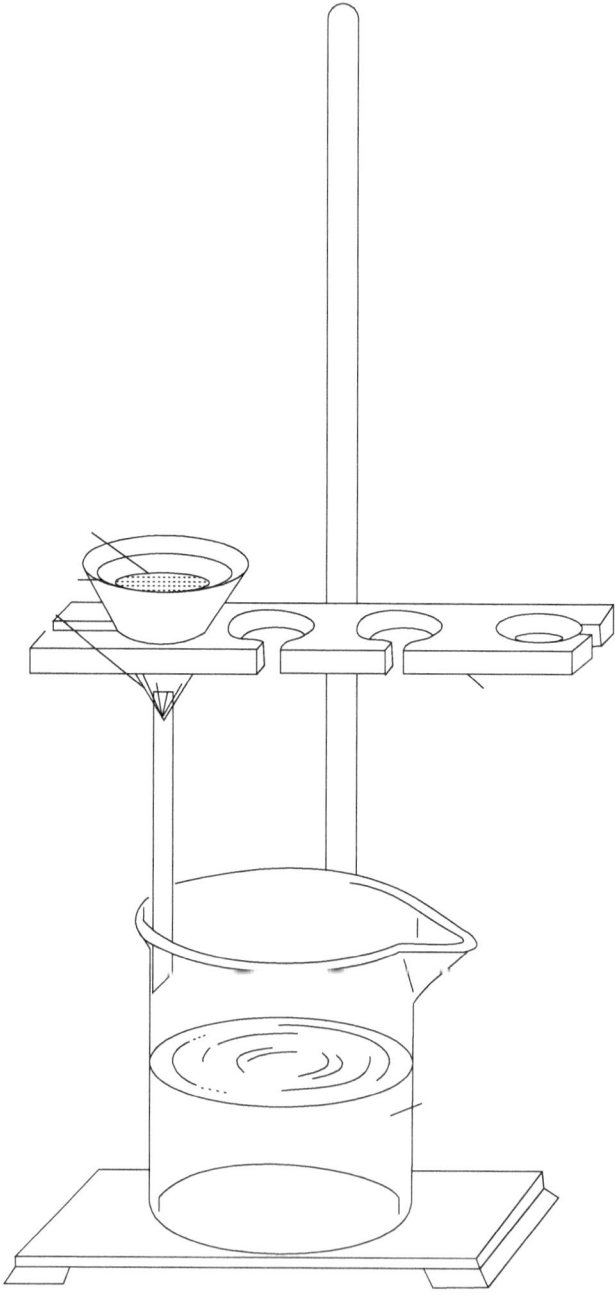

Figure 51. Filtration Apparatus
(**Source:** *Chemical Technician's Ready Reference Handbook*, 3rd ed., McGraw-Hill, 1990, p. 214)

Note the differences in solubility of each gum in different solvents. The best solvent is the ethanol which seems to put all gums into solution, The hydrochloric acid worked some with the agar but didn't do very well with the other gums. Comparing the relative solubilities of the gums karaya was the most readily soluble in all the solvents including water.

This information, compiled on the basis of laboratory research, affects the methods of compounding medicines either in individual compounding pharmacies which are increasing in number and also the large scale manufacturing of drug products.

GUM TRAGACANTH

BOTANY – The botanical name for Tragacanth is *Astragalus gummifer*, a low-growing shrub which gets no more than 30 cm high. The shrub has gray branches which become glabrous. The 8-14 leaflets are folded, oblong-ovate, 2.5-6 mm long and 0.7-2.5 mm wide, blue-gray and glabrous or sparsely pubescent beneath. The medicinal parts are the gum-like exudations from the stems. The plant grows in Turkey, Syria, Lebanon, and northwest Iraq.

CHEMISTRY – Composition is mainly polysaccharides which are water soluble, approximately 40% tragacanthine and other non water-soluble polysaccharides.

Test: Add 1 gm of tragacanth to 50 ml of water. It swells up and forms a smooth, uniform, stiff opalescent mucilage free from cellular fragments.

ADDITIONAL botanicals which exhibit demulcent properties include aloes, glycyrrhiza, and gum karaya.

DERMATOLOGICALS

PHARMACY – Dermatology is the branch of medicine which treats diseases of the largest body organ—the skin. Botanicals used on the skin comprise a wide range of activity including:

- ☐ emollients which soften the skin
- ☐ astringents which cause contraction of the skin
- ☐ rubefacients which induce hyperemia
- ☐ vesicants which are more severe irritants
- ☐ demulcents which protect skin and relieve irritation
- ☐ schlerosing which promote fibrosis

PODOPHYLLUM

BOTANY – *Podophyllum peltatum*, or Mayapple, is a perennial of about 40 cm tall. It has a bifurcated 45 cm high stem and deeply-indented, umbrella-like, hand-sized leaves. The rhizome is reddish-brown and is 0.5 cm in diameter. The solitary white flowers are located in the stem bifurcation between two leaves. The medicinal parts are the dried rhizome and the resin extracted from them. The plant is indigenous to the northeast United States.

CHEMISTRY – Chief components are podophyllotoxin (composing about 20%) plus other lignans including alpha-peltatin, beta-peltatin, and dioxypodophyllotoxin.

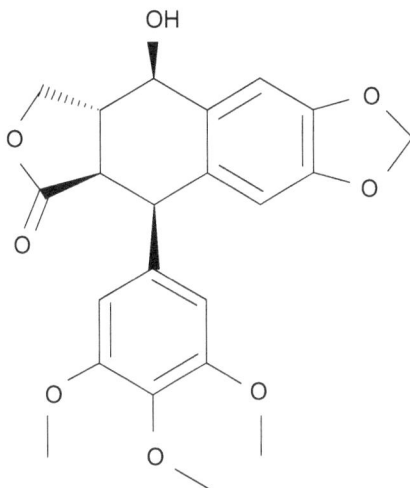

Figure 52. Podophyllotoxin molecule

<u>Assay</u>: Place 10 gm of Podophyllum powder in a 250 ml Erlenmeyer flask and add 35 ml of ethanol. Fit the flask with a reflux condenser and heat on a water bath for three hours. Then transfer the mixture to a small percolator and percolate slowly with warm ethanol until the percolate measures 95 ml. Cool, add sufficient ethanol to make 100 ml and mix thoroughly.

Transfer 10 ml of this percolate to a separatory funnel and add 10 ml of chloroform and 10 ml of 1% hydrochloric acid. Shake the mixture and allow it to separate. Draw off the alcohol-chloroformic mixture into a second separatory funnel and wash the acid layer 3 times with an ethanol-chloroform mixture of 1:2 volume ratio. Add 10 ml of 1% hydrochloric acid to the combined extract and washings. Shake the mixture again and allow to separate, drawing off the alcohol-chloroformic, adding the washing to a tared vessel.

Evaporate the combined extractions on a water bath to apparent dryness. Add 1 ml of dehydrated ethanol and evaporate to dryness.

Weigh the residue which represents the amount of resin in the original sample of Podophyllum. It should read no higher than 4%.

CHRYSAROBIN

BOTANY – Chrysarobin is another name for purified Goa powder which comes from the Andira Araroba tree. It is a large tree whose yellowish wood has vertically running channels and spaces. The latex collects increasingly in these spaces as the tree ages. The bark forms in long flat pieces about 3 mm thick and is grayish-white and fissured externally. The tree is native to Brazil. The medicinal part is the dried and pulverized latex of the trunk and branches.

CHEMISTRY – Composed mainly of anthrone derivatives such as Chrysophano-lanthrone.

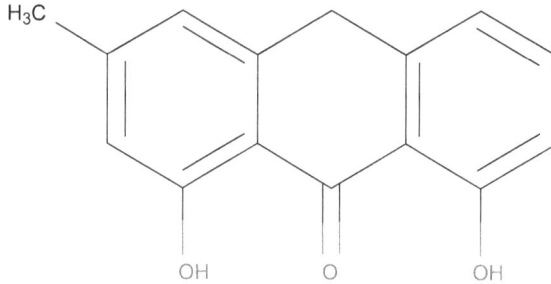

Figure 53. Chrysarobin molecule

Test: Dissolve 1 gm of Chrysarobin in 10 ml of 1.0 M sodium hydroxide. This will result in a deep red color. Repeat the procedure using dilute sulfuric acid. This also yields a deep red color.

Test: Mix 0.1 gm of Chrysarobin with 2 ml of fuming nitric acid. This yields a reddish-brown color. Now add an ml of ammonium hydroxide and note the violet red color.

CHAULMOOGRA

BOTANY – The seeds of the *Hydnocarpus species* are 2-3 cm long and 1.5 cm in diameter. The kernel is oily and endorses two thin heart-shaped three-veined cotyledons. The plant is native to Malaysia and the Indian subcontinent.

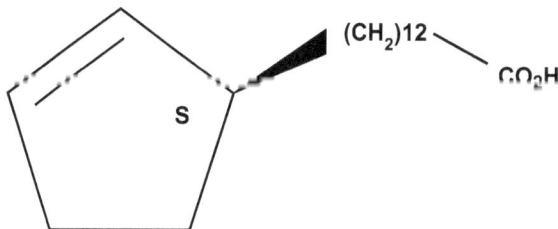

Figure 54. Chaulmoogric Acid molecule

CHEMISTRY – Compounds are mainly fatty acids such as D-Chaulmoogric acid and D-Hydnocarpic acid plus other cyclopentene fatty acids and cyanogenic glycosides.

JOJOBA

BOTANY – Known as *Simmondsia chinesis*, the plant is a heavily-branched ever-green dioecious bush. The desert variety develops taproots up to 3.6 meters in length. The horizontal root branches reach from 60-90 cm in depth. The leaves are thick, coriaceous, blue-green, entire margined, and oblong. The plant is indigenous to areas extending from the Sonora desert in the United States to northwest Mexico and is also cultivated in India and Israel.

CHEMISTRY – Composed of liquid wax esters in position 9-10 of simple unsaturated C-20 and C-22 fatty acids, chiefly gadolenic acid.

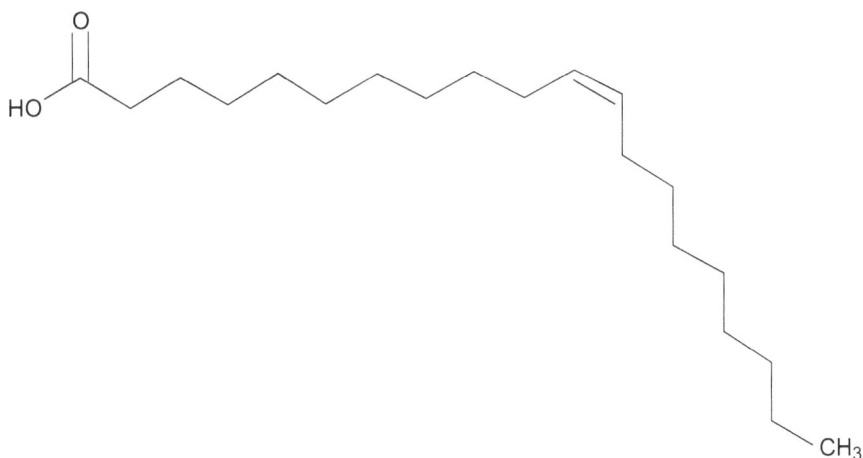

Figure 55. Gandolenic Acid molecule

AGRIMONY

BOTANY – *Acrimonia eupatoria* is a plant which reaches 50-100 cm high with an erect stem, villous leaves alternate, irregularly pinnate, and deeply serrate leaflets. The medicinal part is the flowering plant which is cut a few fingers width above the ground and dried. It is indigenous to middle and northern Europe, temperate Asia, and North America.

CHEMISTRY – Composed mainly of catechin tannins.

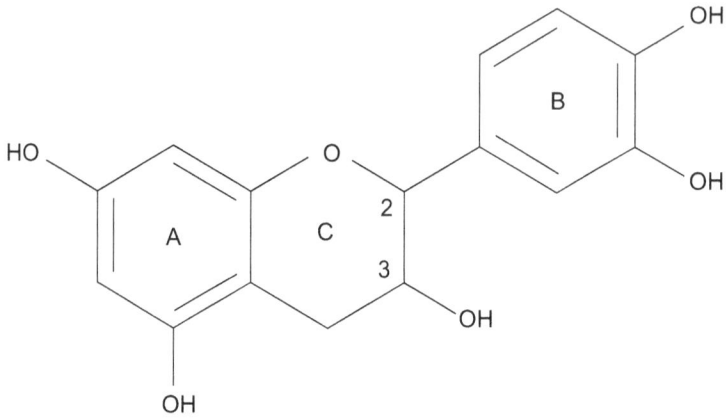

Figure 56. D-Catechin molecule

Test: Boil some Agrimony powder in a small beaker and add a few drops of ferric chloride TS. Note the green color of the solution which is indicative of the catechin tannins as opposed to the pyrogallol tannins which give a blue color with ferric salts.

MARIGOLD

BOTANY – *Calendula officinalis* is usually an annual but occasional perennial that grows between 30-50 cm high with 20 cm long taproot and numerous thin secondary roots. The stem is erect angular, downy and branched from the base up to the top. Its habitat includes central and southern Europe, western Asia and the United States. The medicinal parts of the Marigold are the flowers and the dried above ground part.

CHEMISTRY – Components include *triterpene saponins*, flavenoids, hydroxy-courmarins including scopoletin, carotenoids, and volatile oils.

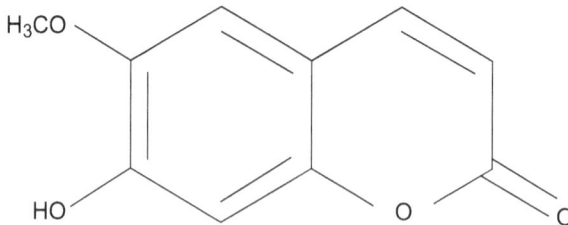

Figure 57. Scopoletin molecule

Tincture of Calendula: Prepare by macerating 20 gm of powder in 100 ml of ethanol for 24 hours. Then percolate slowly.

OLIVE OIL

BOTANY – The Olive, known as *Olea europaea*, is a tree or medium shrub that grows to 10 meters high and has pale bark and cane-1ike quadrangular to round initially downy, thorny, or thornless branches. The leaves are opposite, entire, stiff, coriaceous, narrow elliptical to lanceolate or chorciate thorny tips. The cirupe has 1-2 seeds, is fleshy, plum-like or round, smooth, initially green then red, and is blue black when ripe. The medicinal parts are the dried leaves, the oil extracted from the ripe drupes and the fresh branches with leaves and clusters of flowers.

CHEMISTRY – Composed of iridoide monoterpenes including oleoropine, ligstroside, oleoroside plus triterpenes including oleanolic acid and maslinic acid plus several flavenoids.

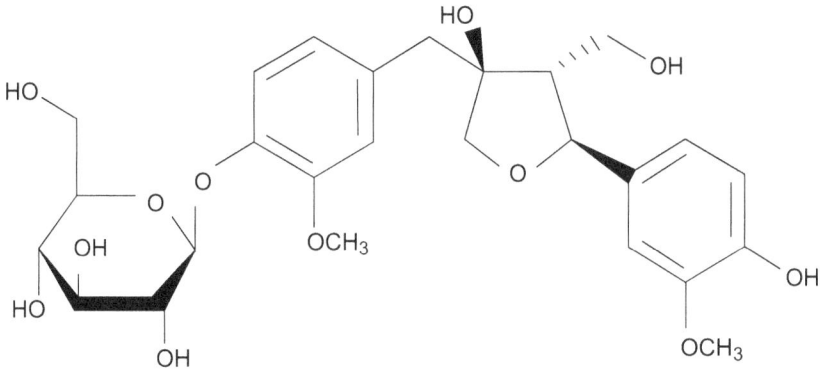

Figure 58. Olivil molecule

Test: For the Iodine Value of olive oil, introduce 0.3 gm of oil into a glass-stoppered flask of 250 ml. Dissolve it in 10 ml of chloroform and 25 ml or iodobromide TS and allow to stand in a cool place for 30 minutes. Now add 30 ml of potassium iodide TS and 100 ml of distilled water. Titrate the liberated iodine with 0.1 M sodium thiosulfate. When the iodine color becomes pale add 1 ml of starch TS and continue titration with the thiosulfate until the blue color is discharged. Carry out a blank test at the same time with the same quantities of chloroform and iodobromide solution allowing it to stand the same length of time and titrating as directed. The difference in the number of ml of thiosulfate consumed by the blank test and the actual test multiplied by 1.269 and divided by the weight of the sample taken gives the Iodine Value. For Olive Oil this should range from 79-88.

PIMENTO

BOTANY – *Pimenta racemosa* is an evergreen tree which grows up to 12 meters high. The leaves are oblong and coriaceous. The fruit is a brown globular berry about 0.75 cm in diameter. This supplies the medicinal oil. The tree is indigenous to the West Indies, but it is also cultivated in Central America, South America, and Jamaica.

CHEMISTRY – The chief components are eugenol (50-60%) and chavicol (20%) plus methyl ether of eugenol, methyl chavicol, and limonene.

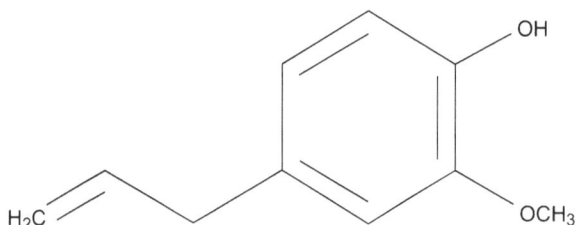

Figure 59. Eugenol molecule

Test: Dissolve 1 ml of Eugenol in 20 ml of 0.5 M sodium hydroxide in a 50 ml stoppered tube as shown in the diagram. Add 18 ml of distilled and mix. A clear mixture results as indicative of hydrocarbon bonds.

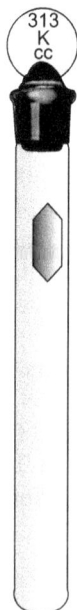

Figure 60. Glass-Stoppered Test Tube
(**Source:** SKS Bottle and Packaging, Inc.,
https://www.sks-science.com/lab-supply-p-8581.html)

Test: Shake 1 ml of Eugenol with 20 ml distilled water. Filter and add a drop of ferric chloride TS. The mixture exhibits a transient grayish-green color which is indicative of the presence of a phenol.

ADDITIONAL botanical materials that possess therapeutic answers for dermatological problems include Myrrh, Peruvian Balsam, Arnica, Capsicum, Cotton Seed Oil, Camphor, Tincture of Benzoin, and Cloves.

ALMOND OIL

BOTANY – *Prunus almygdalus* is the botanical name for the Almond. The plant is a medium high tree or shrub with mildly red-tinged thorny branches. The leaves have a 1.2-1.5 cm long glandular petiole and glabrous oblong-lanceolate-acuminate, finely dentated, glossy, dark green leaf blades. The flowers are very short petioled in pairs and appear before the leaves. The petals are 19-20 mm long, pale pink to white with dark veins. The fruit is oblong-ovoid, compressed, and 3.5 -4.6 cm long. The nut shell is yellow and the seed is common brown. The medicinal part is the ripe fruit. The tree is indigenous to western Asia and is extensively cultivated in other areas.

CHEMISTRY – Compounds are non-dehydrating fatty acids, benzaldehyde plus fatty acids including oleic and linolenic as well as mucilages.

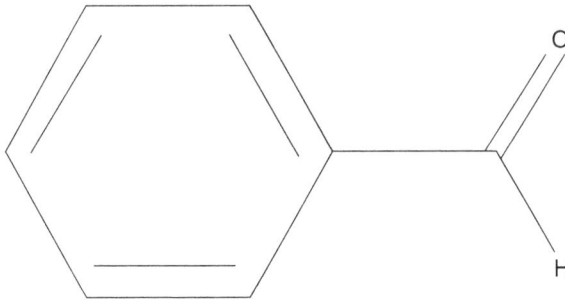

Figure 61. Benzaldehyde molecule

Assay for Benzaldehyde: Transfer 1 ml of Bitter Almond Oil to a tared glass-stoppered weighing bottle and weigh. Transfer the weighing bottle and contents to a 250 ml flask containing 25 ml of a reagent of hydroxylamine in ethanol TS. Allow to stand for 10 minutes add 1 ml of bromphenol blue TS and titrate the liberated hydrochloric acid to a light green end point with 1.0 M sodium hydroxide. Run a blank using the same quantity of reagents but omitting the Bitter Almond Oil. Match the color of this end point in the above assay. To determine the required amount of 1.0 M sodium hydroxide to neutralize the liberated hydrochloric acid, subtract the number of ml of the 1.0M sodium hydroxide used in the blank from the number of ml required in the assay. Each ml of 1.0 M sodium hydroxide is equivalent to 106.1 mg of Benzaldehyde.

DIURETICS

PHARMACY – Diuretics are drugs used to increase the volume of urine excreted by the kidneys. They are employed principally for the relief of edema and ascites. Some diuretics have highly specialized uses as in glaucoma treatment, hyperkalemia, anginal syndrome, epilepsy migraine, and hypertension. Since the glomerular filtration rate is about 100 ml per minute and about 99 ml are returned to the blood without being extracted as urine, diuretics work by increasing glomerular filtration and by depressing tubular reabsorption.

BURDOCK

BOTANY – The botanical name for burdock is *Arctium lappa* which is a plant that grows up to a height of 80-150 cm high. The stem is erect, rigid, grooved, branched, and downy to wooly. The leaves are alternate petiolate broad to ovate-cordate. The lower leaves are very large and have a latex-filled stem. The plant has crimson flowers which growth in long peduncled loose cymes. The flowers are funnel-shaped and androgynous. The medicinal part is the fresh or dried root. Burdock grows in Europe, north Asia, and North America.

CHEMISTRY – Compounds include volatile oils and some aldehydes including phenylacetaldehyde and benzaldehyde, plus sesquiterpenelactones, caffeic acid derivatives, and polysaccharides.

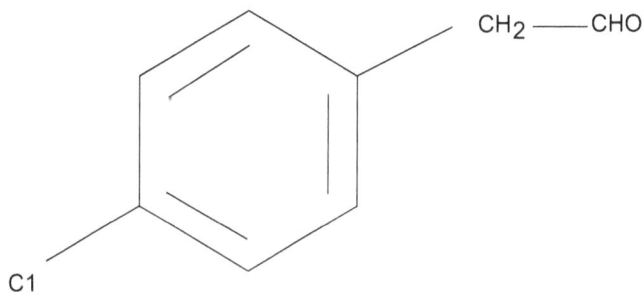

Figure 61. Phenylacetaldehyde molecule

<u>Tollens' Test</u>: Place 2 ml of Phenylacetaldehyde in a test tube and add 2 ml of a 5% solution of silver nitrate and a few drops of 0.1 M sodium hydroxide solution. Then add, drop by drop, a dilute solution of ammonium hydroxide solution while constantly shaking. Some elemental silver will begin to show in the solution as evidence of the presence of an aldehyde. $RCHO + Ag_2O \rightarrow 2Ag + RCOOH$

<u>Schiff's Test</u>: Place 2 ml of Phenylacetaldehyde in a test tube and add a few drops of Schiff's Reagent. A brilliant magenta appears as proof of an aldehyde.

LARKSPUR

BOTANY – The *Delphinium consolida* also known as Larkspur is a plant that grows from 15-40 cm high and has a thin stem which is sparsely branched from the middle. The leaves are alternate and divided into narrow linear sections. The lower ones are petioled and the upper ones sessile. The flowers are in short racemes and are blue, pink, or purple. The medicinal part is flower. The plant grows in Europe and in the western United States.

CHEMISTRY – The main components of Larkspur are diterpene alkaloids, particularly Delphinine.

Formula: $C_{33}H_{45}NO_9$, mol wt = 599.7, melt pt = 197.5-199

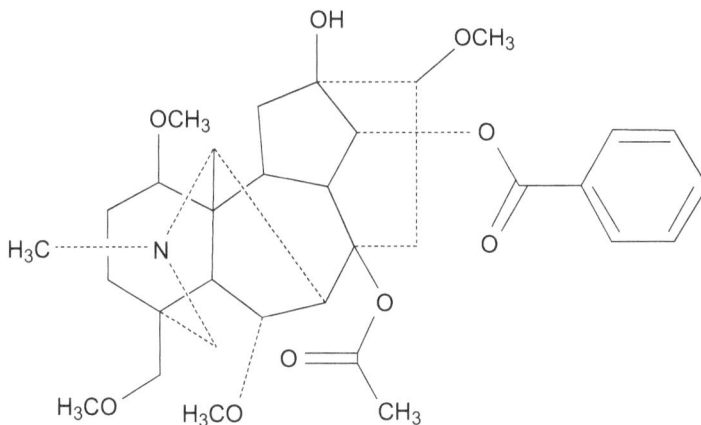

Figure 62. Delphinine molecule

<u>Determining the melting point of Delphinine</u>: Set up an apparatus using a round- bottom tube of 30-40 cm diameter. Place the drug in a capillary tube and attach to the thermometer using platinum wire. Adjust the height so that the powdered drug is as close to the stem of the thermometer as possible at a point midway between the surface of the bath and the graduation for the supposed melting point. Heat the water bath until a temperature of about 25 degrees below the suspected melting point is reached. Now carefully regulate the rate of

rise in temperature to about 3 degrees per minute until the material in the capillary tube begins to melt. The temperature at which the material becomes liquid throughout signifies the end of melting. Depending upon the purity of the sample Delphinine should melt between 197.5 and 199 °C.

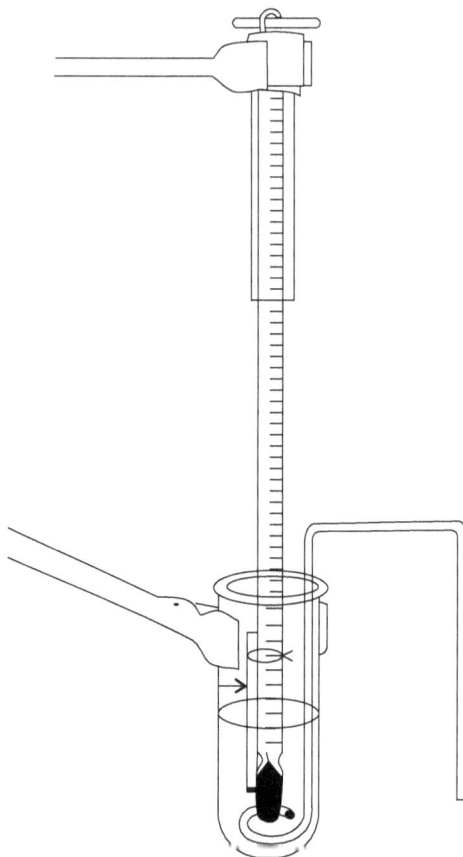

Figure 63. Melting Point Apparatus
(**Source:** *US Pharmacopeia*, 11th ed., Mack Printing, 1936, p. 456)

CLUB MOSS

BOTANY – *Lycopodium clavatum*, or Club Moss, is a plant 1 meter long with a procumbent stem with only a few roots. It is covered with yellowish-green leaves densely arranged in spirals which are entire-margined, linear, smooth, and end in a long white, upwardly-bent, hair tip. The medicinal parts are the spores and the fruit. The plant is found worldwide.

CHEMISTRY – Components are mainly alkaloids including lycopodine and dihy-drolycopodine. There are also some traces of nicotine plus flavonoids including chrysoerial and luteolin.

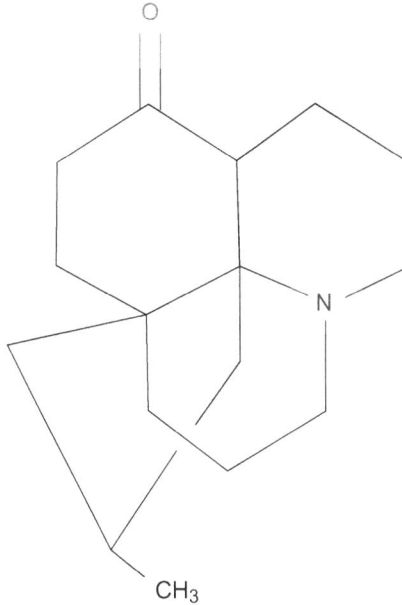

Figure 64. Lycopodine molecule

Test: Place some Lycopodium powder on top of water in a beaker and note that the powder is not wetted by the water. Only when the solution is boiled does the powder sink.

ANGELICA

BOTANY – *Angelica archangelica* grows from 50-250 cm. The rhizome is short, strong, thick, fleshy, and has long fibrous roots. The stem is erect, thick, round, finely-grooved, hollow, and tinged reddish-brown below the medicinal parts are the seed, whole herb, and root. It is thought to be indigenous to Syria but can be discovered growing wild on the coasts of the North and Baltic Seas and as far north as Lapland. It is also cultivated in other regions of the world including the United States, Europe, and China.

CHEMISTRY – Compound include the volatile oils such as alpha-Phellandrine, beta-Phellandrene, alpha-Pinene, macrcyclic lactones including penta- and hepta-decanolide. There are also furocoumarins including beragatiene, xantho-toxin, and umbelliferone plus numerous flavanones.

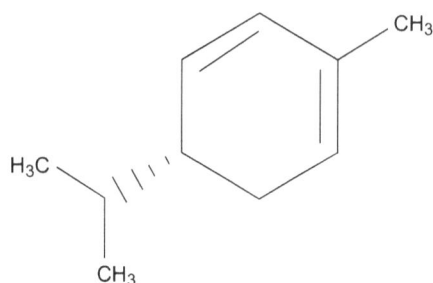

Figure 65. Alpha-Phellandrene molecule

DANDELION

BOTANY – The Dandelion, *Taraxacum officinale*, is a perennial plant, hardy and found in a number of forms. It grows to about 30 cm and has a short rhizome. The hollows stem is erect or ascending. The flower is a golden yellow composite. The composite head is solitary and has a diameter of 3-5 cm. The medicinal parts are the dried leaves harvested before the flowering season and the dried roots in autumn. The Dandelion grows in most temperate regions of the world.

CHEMISTRY – Components include sesquiterpene lactones such as taraxacolide and triterpenes such as taraxasterol.

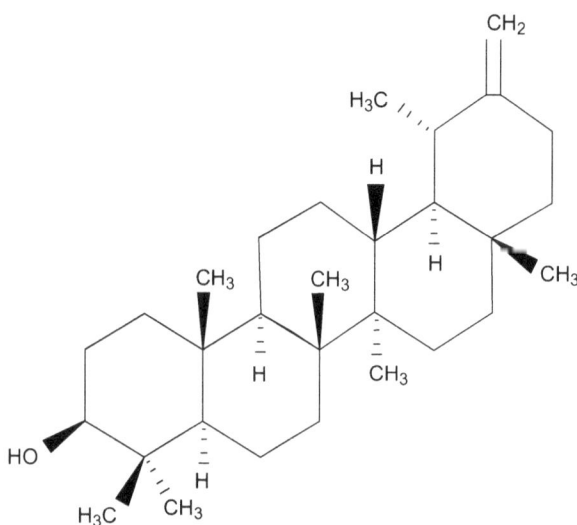

Figure 66. Taraxasterol molecule

ADDITIONAL botanicals with diuretic action include Sweet Clover, Garden Cress, Hibiscus, Hampnettle, Goat's Rue, and Meadowsweet.

EMETICS

PHARMACY – Emetics are drugs which induce vomiting. Although vomiting is primarily a respiratory function, the final reaction is the evacuation of the stomach. Some drugs may act directly by their stimulation of the chemoreceptor trigger zone or they may act reflexly by their irritant action on the gastrointestinal tract. They may also produce stimulation of the nodose ganglion of the heart in the central nervous system rostral to the brain stem and other organs.

Over time the emetic drugs have saved many lives by stimulating the evacuation of stomach contents and thereby reducing absorption of poisonous substances.

IPECAC

BOTANY – *Cephalaelis ipececuanha* is a perennial evergreen leaf plant about 40 cm high with a 2 -4 mm thick rhizome from which sprout numerous 20 cm long fibrous roots some of which develop into tubers. The leaves of the plant are opposite and entire margined and the leaf blade narrows into the short petiole. The medicinal parts are the pulverized ipecacuanha root of the 3-4 year old plant which have been dug up and dried in the sun. Ipecac is indigenous to the sparser wood of Brazil and also cultivated in India.

CHEMISTRY – The principal active ingredient is the alkaloid emetine in addition to the alkaloid, cephaelin.

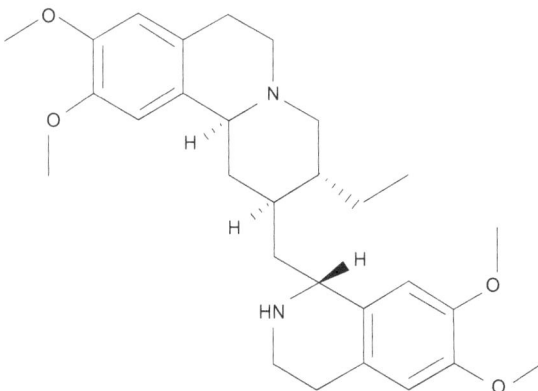

Figure 67. Emetine molecule

Assay: In a 250 ml Erlenmeyer flask, place 3 gm of Ipecac powder in 25 ml of ether. Add an excess of sodium hydroxide solution and shake the mixture for several minutes. Separate the ether layer using a separatory funnel and complete the extraction of the alkaloids (Emetine and Gephaelin) with successive portions of ether. Combine the ether solutions and extract the alkaloids completely using dilute sulfuric acid (2%), keeping the volume of the acid as small as is practical. Filter them successively through a plug of cotton. Next render the combined acid solutions alkaline with a short excess of ammonia TS and extract the alkaloids completely by shaking with successive portions of ether.

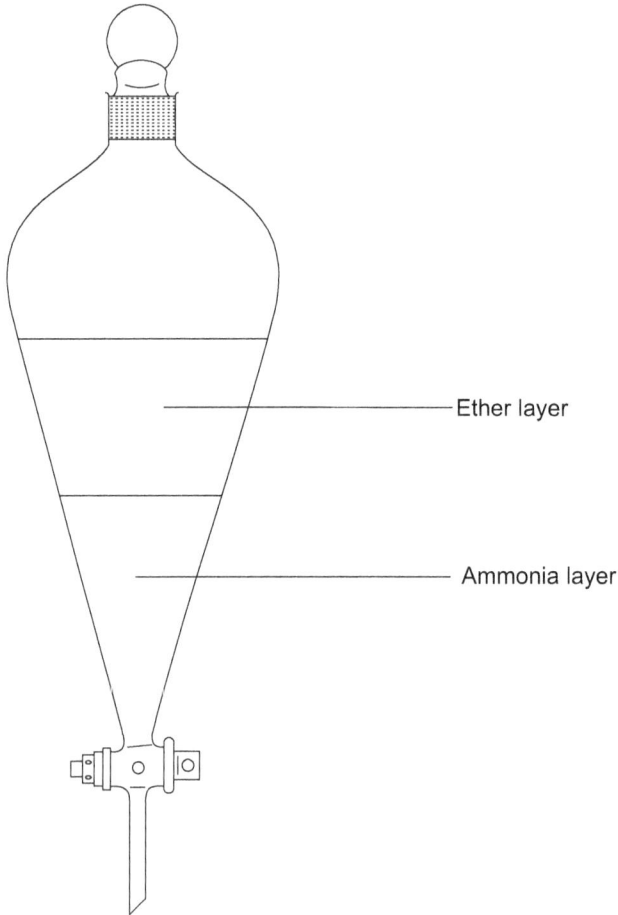

Figure 68. Separatory Funnel showing fluids

Now evaporate the combined ether solutions nearly to dryness and add 10 ml of 0.1 M sulfuric acid. Warm to dissolve the alkaloids and to expel the remaining ether. Cool and titrate the excess acid 1with .02 M sodium hydroxide using methyl red TS as the indicator.

Each ml of the 0.1 M sulfuric acid used is equivalent to 24.0 mg of the ether-soluble alkaloids (Emetine and Cephaelin) of Ipecac.

BLACK MUSTARD

BOTANY – *Brassica nigra*, Black Mustard, is an annual tall-growing, slim-branched plant with thin fusiform roots. It is almost round and bristly haired at the base with inflorescence as terminal axillary and compressed into a semi-space. The flowers are yellowish-green. The medicinal part is the seeds. Oil is present in the seeds. Black Mustard is a plant that is distributed world-wide in all the temperate regions.

CHEMISTRY – Allyl Isothiocyanate (3 isothiocyanato 1-propane) is a fatty oil which comprises 30-35% of the seeds. They are also about 40% protein plus phenyl propane derivatives.

Figure 69. Allyl Isothiocyanate molecule

ADDITIONAL botanicals which possess emetic properties are Acalypha and Poke Weed.

EXPECTORANTS

PHARMACY – Expectorants are drugs that are used to assist the removal of secretion or exudate from the trachea bronchi, or lungs and hence they are useful in the treatment of cough. Expectorants may also have anesthetic, antiseptic, or diuretic use. Expectorants are usually divided into categories such as sedative expectorants, stimulant or irritative expectorants, and centrally acting expectorants, and antitussive agents.

ANISE

BOTANY – Anise, whose botanical name is *Pimpinella anisum*, is an annual herb about 0.5 meters high and downy all over. The root is thin and fusiform and the stern is erect, round, grooved, and branched above. The lower leaves are petiole, or orbicular-reniform entire, and coarsely dentate to lobed. The middle leaves are short-petioled to sessile. The medicinal parts are the essential oil from the ripe fruit. The origin of the plant is unknown but it probably came from the Middle East. Today it is cultivated in Europe, Turkey, Central Asia, India, China, Japan, Central and South America.

CHEMISTRY – The chief constituent is the volatile oil, trans-anethole. Present also are chavicol and anisaldehyde.

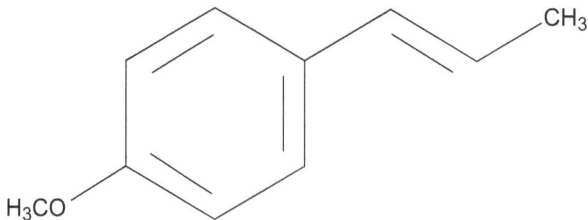

Figure 70. Anethole molecule

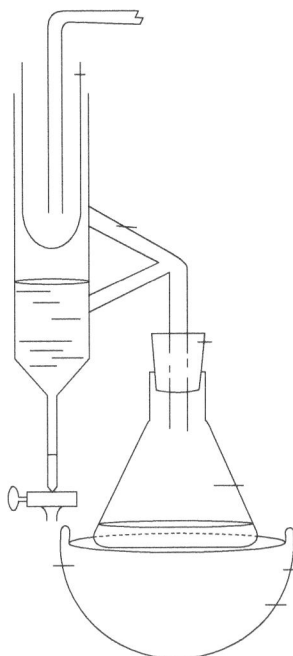

Figure 71. Volatile Oil Determination Apparatus

<u>Assay for Volatile Oil Determination</u>: Place 50 gm of Anise in a flask with 100 ml of ethanol and condenser connected to a volatile oil collector as shown. Heat using a heating mantle for 2 hours. Continue until all the oil has been separated from the drug and collected. Now measure the amount and com pare to the original. The percentage yield of volatile should average from 1.5-2.0%.

SOAP BARK

BOTANY – *Quillaja saponaria*, or Soap Bark, is a tree which grows up to 18 meters tall. The leaves are smooth, glossy, short-petioled, and oval. The bark is thick, dark, and very hard. The terminal inflorescence consists of white androgynous flowers with a calyx and corolla but no epicalyx. Three to five flowers are grouped on the peduncle. The seeds are winged with little or no endosperm. The medicinal part is the inner bark. The tree is indigenous to Chile and Peru, but is also cultivated in India and in California.

CHEMISTRY – Compound consist mainly of triterpene saponins, chief of which are quillaja saponina, particularly quillaic acid. The saponins range up to 17% of the botanical material. There are also tannins present which vary from 10-15%.

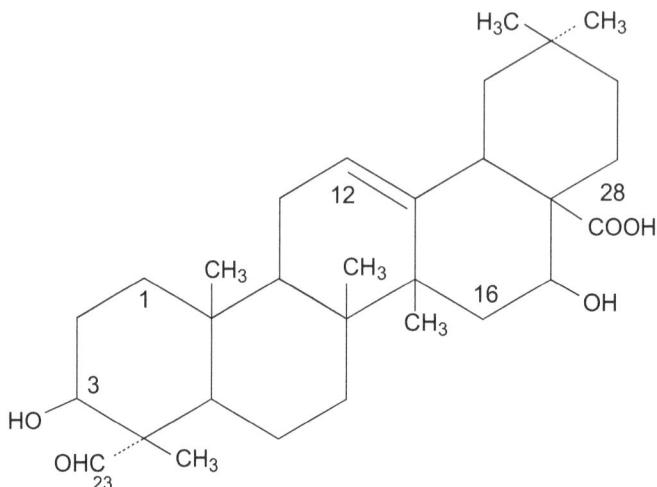

Figure 72. Quillaic Acid molecule

<u>Acid-Insoluble Ash Test</u>: Place 4 gm of the Quillaja bark powder in a tared crucible and incinerate at a low temperature until free from carbon. Then determine the weight of the ash. If it is not, carbon-free, exhaust the charred mass with hot water and collect the insoluble residue on filter paper. Add some ethanol, then burn in off and calculate the total ash. Next boil the ash in 25 ml of hydrochloric acid for 5 minutes, collect the insoluble, wash with hot water, ignite, and weigh. The acid-insoluble ash of Quillaja should be about 2%.

<u>Saponin Content Test</u>: Boil 20 gm of bark in 100 ml of water. Collect and dry the filtrate and place it in 50 ml of hydrochloric acid. Boil under a reflux condenser for an hour. Filter dry and weigh the insoluble sapogenins. One part of sapogenin is equal to 3.22 parts saponin.

SENEGA

BOTANY – *Polygala senega* is a perennial herb with up to 40 cm high stems which sprout in the axis of the scale-like bracts of the previous year's growth. The leaves are 8 cm long and 3 cm wide, alternate, ovatelanceolate to lanceolate, acuminate, and denticulate. The petals are pale red. The wings are yellowish-white with green veins. The medicinal part is the dried root. The plant is indigenous to central and western United States.

CHEMISTRY – Components include triterpene saponins including senegin (6-12%) xantho derivatives and oligosaccharides.

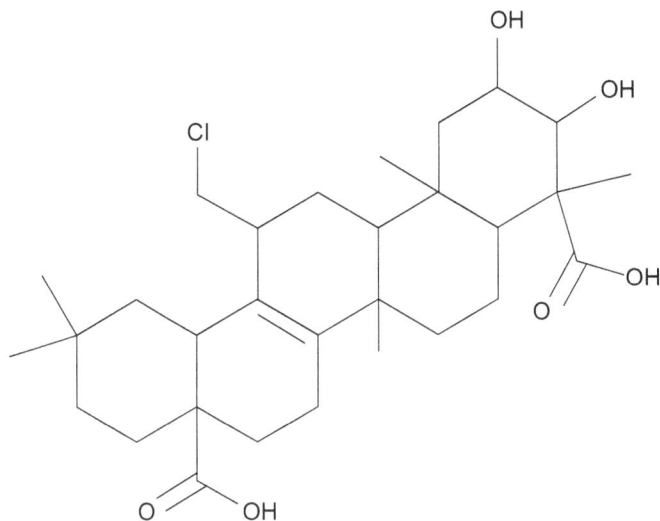

Figure 73. Senegenin molecule

Preparation: Prepare Fluidextract of Senega by placing 3 gm of powdered root in 100 ml of an ethanol:water solution (2:1). Macerate the drug for 48 hours, then percolate at a moderate rate.

PERUVIAN BALSAM

BOTANY – *Hyroxylon balsamum* is a tree that grows up to 25 meters tall. The bark is yellowish-gray or brown, with numerous lenticels. The leaves are usu- ally odd-pinnate and have 4-7 obovate, acuminate coreaceous short-petioled leaflets. The medicinal parts are the balsam from the sweltered trunks. The tree is indigenous to Central and South America.

CHEMISTRY – Esters, chiefly benzyl benzoate and benzyl cinnamoate, make up 50-70% of the compound. Resins make up another 20-30%.

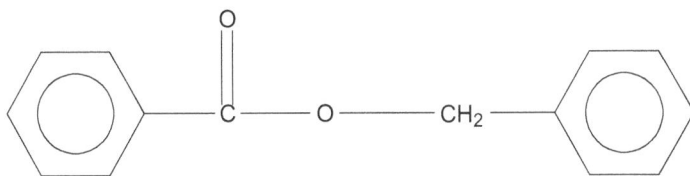

Figure 74. Benzyl Benzoate molecule

Assay: Place 25 ml of saponated benzyl benzoate in a 100 ml flask and dilute to the mark with ethanol. Transfer 5 ml of this dilution to an Erlenmeyer flask. Add 2 drops of phenolphthalein TS and titrate with alcoholic potassium hydroxide.

Connect to a condenser reflux for an hour, cool and titrate with 0.5 M hydrochloric acid using phenophthalein as test indicator. Perform a blank test using same procedure and compare. Each ml of 0.5 M potassium hydroxide is equivalent to 106.1 mg of benzyl benzoate.

MYRRH

BOTANY – *Commiphora momol*, known as Myrrh is a stunted shrub growing to 3 meters high with a thick trunk and numerous irregular knotted branches and smaller clustered branchlets. It has a few trifoliate leaves at the end of short branches. The medicinal part is the resin exuding from the bark. The shrub is indigenous to the eastern Mediterranean and Somalia.

CHEMISTRY – Composed of 2-10% volatile oils including sesquiterpenes, furosesquiterpenes including 5-acetoxy 2-methoxy,4,5 dienone which is the aroma bearer. It also has triterpenes, including 3 epi-alpha amyrin and mucilages, chiefly glucorono-galactans.

Test: Place 2 ml of Myrrh in a test tube and add a few drops of nitric acid. Note the purplish color.

Test: Place 2 ml of Myrrh in a tube and 2 ml of ether. Treat the mixture with bromine vapor and note the reddish-violet color which develops.

Test: Triturate some tincture of Myrrh with water and note the yellowish-brown solution which results.

GUMWEED

BOTANY – *Grindelia camporum* is an erect biennial or perennial bush up to 1 meter high, often branched above. The alternate flowers are 3-7 cm long, triangular to ovate-oblong, clasping, resinous-punctate, serrate-crenate or entire-margined and light green. The medicinal parts are the flowering branches. The plant's habitat is the southwest United States and Mexico.

CHEMISTRY – Composed of diterpene acids including grindelic acid, hydroxygrindelic acid, and 6-oxogrindelic acid. There are also numerous saponins and tannins.

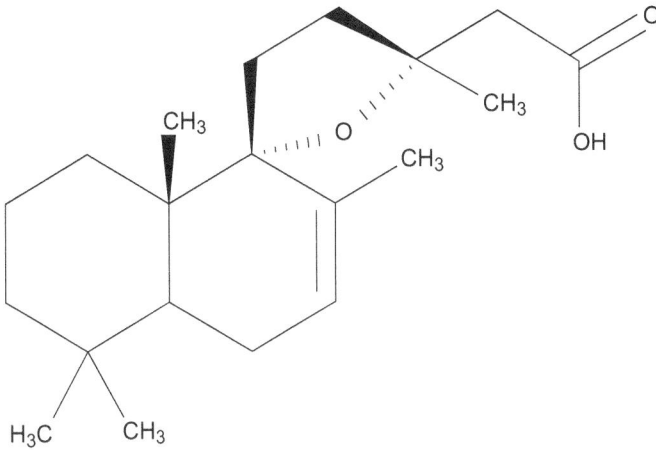

Figure 75. Grindelic Acid molecule

ADDITIONAL botanicals with expectorant properties include Pleurity Weed, Mullein, Quegracho, Licorice, and Cubeb.

GENITO-URINARY

PHARMACY – The term Genitourinary covers a wide range of anatomical subjects and physiological functions including male and female genitalia, the organs and parts concerned with the kidneys, urinary bladder, and organs of reproduction. Part of this subject has been covered under the APHRODISIACS and the DIURETICS listing so this section will deal primarily with the botanical drugs used to treat problems of the genitalia, gynecological, obstetrical, and general reproductive system.

ERGOT

BOTANY – Ergot, known botanically as *Claviceps purpurea*, is a fungus which is a parasite in ripening rye, wheat, and other grasses. It is black, hard, and much larger than the grains of rye. *Claviceps purpurea* is a widespread fungus not limited to any particular global region. The medicinal part is the Ergot in the form of the sclerotum which has formed on the rye and has been dried.

CHEMISTRY – Compounds include alkaloids associated with lysergic acid amide include ergometrine, ergotamine, ergovaline, egocristine, and others plus xanthone derivatives including secalonic acid, anthracene derivatives including clavorubine and endocrocine, amines including methylamine, and trimethylamine, plus fatty oils.

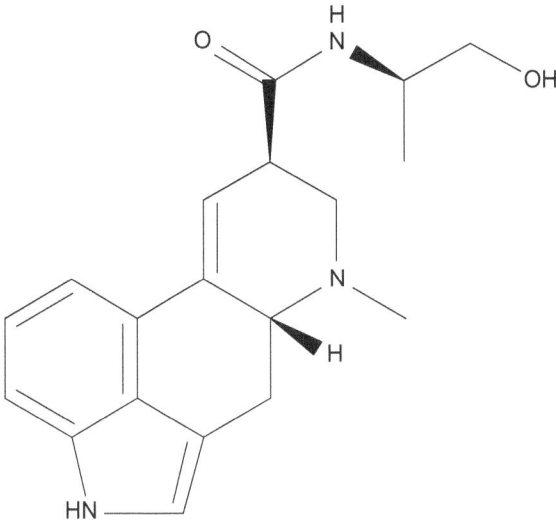

Figure 76. Ergometrine molecule

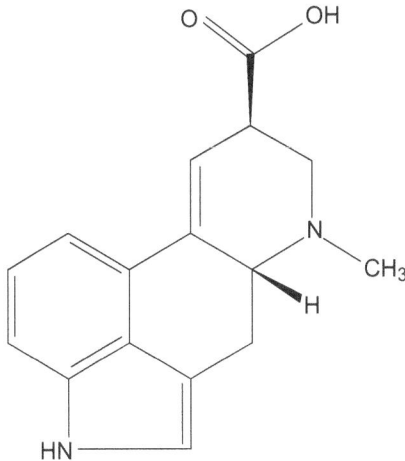

Figure 77. Lysergic Acid molecule

Assay for Ergometrine (Ergonovine): Transfer 20 ml of a concentrated solution of total alkaloids representing 4 gm of Ergot into a 250 ml separator and add 30 ml of distilled water. Render the solution slightly alkaline using ammonia TS and phenolphthalein TS as an indicator. Extract the alkaloid mixture with 3 successive 30 ml portions of carbontetrachloride washing each portion with the same 20 ml portions of distilled water contained in a second 250 ml separator. Discard the carbon tetrachloride. Combine the water solutions and extract the

Ergometrine with ether as follows: Add 4.0 ml of ether and sufficient sodium chloride to saturate the water phase and shake 5 minutes until the sodium chloride is completely dissolved. When settled, draw off the water phase into the second separator. Let stand about 5 minutes longer allowing time for the water adhering to the walls of the separator to collect, then drain this off also. Pour the ether phase from the mouth of the separator through a cotton pledget previously moistened with ether into a third 250 ml separator. Extract the ergometrine from the combined ether solutions with a total of 4 successive 10 ml portions of 0.2 M sulfuric acid. Filter each portion through a small cotton pledget rinsing the pledget with a little water following the last extraction. Dilute the combined water extracts to 50 ml stopper securely and mix. To 10 ml of this solution, add 20 ml of dimethylaminobenzaldehyde TS. Mix thoroughly and allow to stand for 30 minutes. By means of a suitable photometer, determine the optical density at 500 M of the of water solution 1 ml in depth relative to a blank composed of l0 ml of water and 20 ml of p-aminodimethylbenzaldehyde TS. Compare the reading directly with that produced by a solution of Ergotmetrine Reference Standard having a concentration which varies from that of the sample not more than 50%.

Express the result as mg of Ergometrine in each gm of Ergot. Each mg of the reference is equivalent to 0.737 mg of Ergometrine.

BLACK HAW

BOTANY – *Viburnum prunifoilum* is a deciduous tree 5 meters in height. It has gray-brown bark and green-grooved branches. The leaves are opposite, petiolate, 3-5 1obed, roughly dentate, and green on both surfaces. The medicinal parts are the trunk and the root. The habitats of Black Haw are the eastern and central United States.

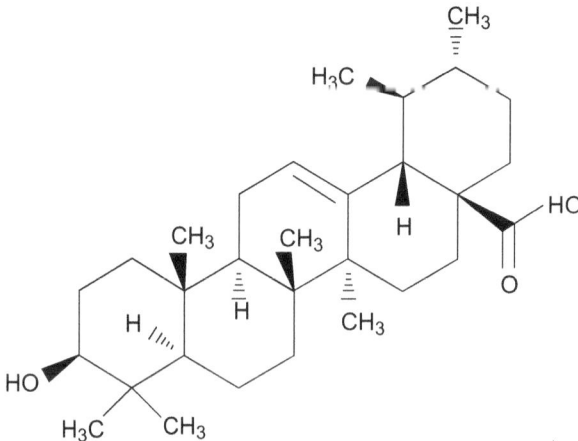

Figure 78. Ursolic Acid molecule

Preparation: Prepare a fluidextract by mixing 10 gm of the drug with about 600 ml of an ethanol: water solution to wet the drug. After 15 minutes, place this mixture in a percolator using an ethanol:water menstruum, adding sufficient solvent. Proceed with percolation, adding fresh menstruum as needed. Reserve the first 850 ml of percolate.

PITCHER PLANT

BOTANY – *Sarracenia purpurea* is a perennial plant whose leaves, which are in a basal rosette that change into a tube or beaker-like formation which bears a long wing-like step on the side turned toward the stem. The beakers are very colorful, fill up with rain water, and trap insects. The medicinal parts are the leaves and roots. It is indigenous to the United States.

CHEMISTRY – Compounds are mainly piperidine alkaloids such as coniine and gamma-conicein.

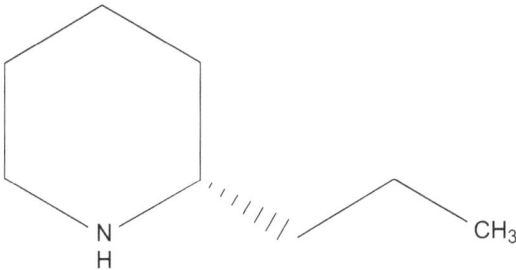

Figure 79. Coniine molecule

Test: Place 0.5 gm of coniine powder in a small beaker and add 2 ml of Dragendorff's Reagent (a mixture of potassium iodide and bismuth iodide).

Test: Repeat the above procedure using iodine TS. Again, the result is a reddish-brown color.

Test: Repeat the above using bromine water. Note the yellow color which develops.

BLACK COHOSH

BOTANY – *Cimicifuga racemosa* is a plant that grows from 1-1.5 meters high. It is leafy with a sturdy, blackish Rhizome which is cylindrical tough and knotty. The transverse root section shows wedge-shaped bundles of white wood, while the section of rhizome shows a large medulla surrounded by a ring of paler woody wedge. The leaves are double-pinnate, smooth, and crenate-serrate. The medicinal part is the fresh and dried root. The plant is indigenous to Canada and the United States but is also cultivated in Europe.

CHEMISTRY – Compounds include triterpene glycosides including actein and 27-deoxyactein, cimigenol and phenylpropane derivative including isoferulic acid.

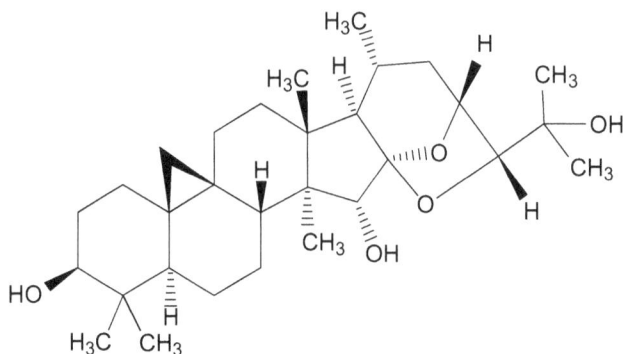

Figure 80. Cimigenol molecule

SASSAFRAS

BOTANY – *Sassafras albidum* is a deciduous tree up to 30 meters high with numerous thin branches. The bark of the trunk and of the thicker branches is rough, deeply grooved, and grayish. The alternate leaves are petiole and 7-12 cm long. Some are simple ovate, others deeply 2 or 3 lobed. The root is brownish white, showing clear concentric rings transversed by narrow medulla rays. The medicinal parts are the essential oil of the root wood, the peeled and dried root bark, and the root wood. Sassafras trees appear in countries throughout the world.

CHEMISTRY – Composed of volatile oils chiefly saffrole, 5-methoxyeugenol, asarone, and camphor plus isoqinoline alkaloids.

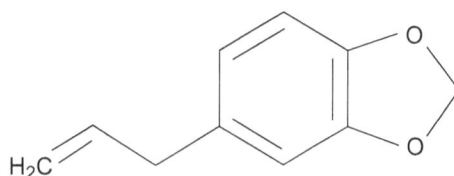

Figure 81. Safrole molecule

Volatile Oil: Place 3 gm of sassafras bark in 100 ml of ethanol. Set up a steam distillation process as shown in the diagram. Begin heating the steam generator and heat until no more volatile oil appears in the receiving flask. The amount of volatile oil collected should average about 9% of the total weight of the Sassafras bark, but there are also asarone, camphor and 5-methoxyeugenol in the mix as well, generally in lesser amounts.

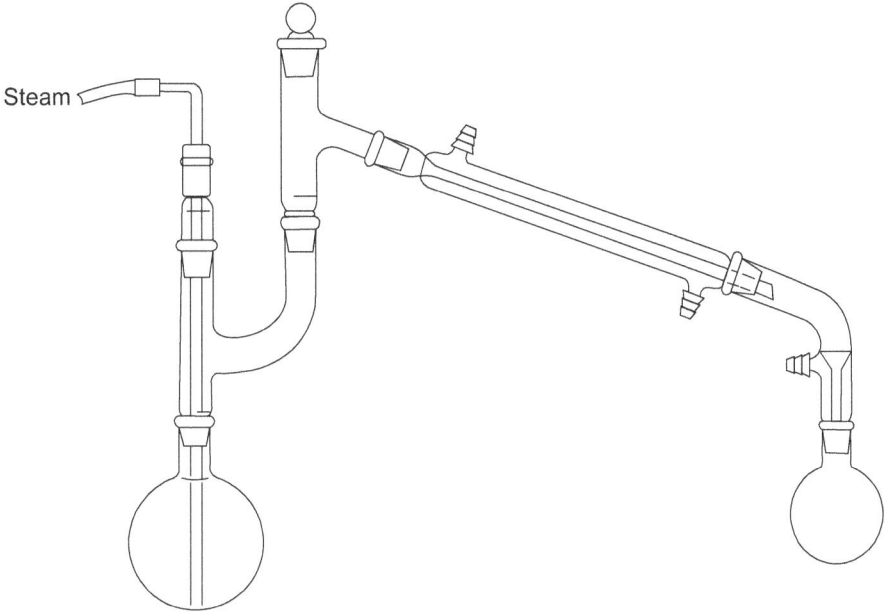

Steam

Figure 82. Distillation Apparatus

ADDITIONAL botanicals which are used to treat the Genito-Urinary system include Broad Bean, Rue, Goldenseal, Buchu, and Juniper.

LAXATIVES

PHARMACY – Laxatives, purgatives, and cathartics are three terms often used interchangeably. However, the latter two may be regarded as having stronger action on the bowels. All three are designed to facilitate the passage and elimination of feces from the colon and rectum. There are subclassifications which focus on site and mode of action. Those referred to as irritants act on the intestinal tract to increase its motor activity and include aloe, cascara sagrada, and rhubarb while the resinous type tends to produce watery stools. Their focus of action is the small intestine. Included in this group are colocynth, jalap, gamboges, and ipomea.

JALAP

BOTANY – *Ipomoea orizabensis*, or Jalap, is a twining plant with large cordate leaves. The root tuber is about 18-25 cm long and 9-10 cm wide and cylindrical-fusiform. It is grayish-brown to brownish-black black and wrinkled externally. The plant has reddish-purple and campanulate flowers. The medicinal parts are the dried roots and steamed ethanol extract from the roots. It is indigenous to Mexico.

CHEMISTRY – The active components are glycoretines or resinous ester glycosides.

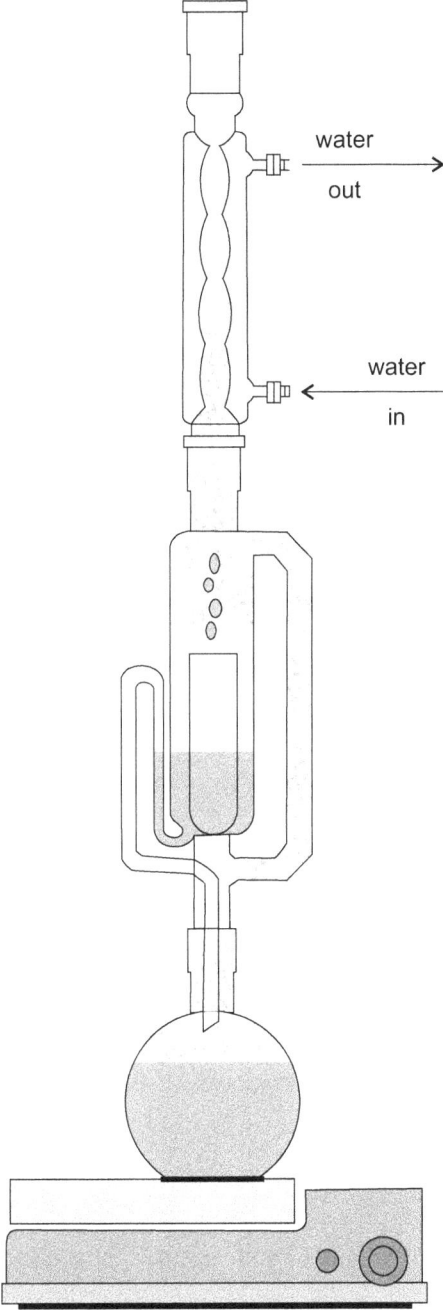

Figure 83. Soxhlet Apparatus
(**Source:** Tomasz Dolinowski illustration, Wikimedia Commons)

Assay: Place 5 gm of Jalap powder in the boiling flask of a Soxhlet Apparatus as shown. Add 100 ml of 90% ethanol. The extraction thimble should be plugged with cotton wool and the boiling process begun. The ethanol is removed by distillation and the concentrated tincture transferred to a dish. Weigh the dish and stirring rod, treat the residue with boiling water, and filter. Repeat this and finally dissolve the residue on the filter paper with hot ethanol which is allowed to filter through into the weighed dish containing the bulk of the pure resins. Evaporate the ethanol by drying at 100 ºC and complete the procedure by weighing again the dish containing the resins along with the stirring rod. Jalap root resins should calculate to be about 12-15% of the total weight of the bark used at the beginning of the assay.

PSYLLIUM

BOTANY – *Plantago afra*, a form of Psyllium, is an annual that is erect with stems up to 60 cm high. The stems ascending pubescent branches with ascending hairs and are more or less glandular above. The leaves are 3-8 cm by 0.1-0.3 cm linear or lanceolate. The medicinal parts are the ripe seeds. The plant is indigenous to the Mediterranean region and eastern Asia but is also cultivated in other parts of the world.

CHEMISTRY – Compounds are primarily mucilages, chiefly arabinoxylans.

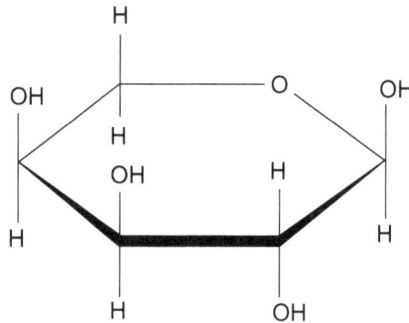

Arabinose

Figure 84. Arabinose molecule

Test: Place 1 gm of Plantago seed in a 25 ml graduated cylinder. Add 10 ml of water. Shake the cylinder at intervals during a 24-hour period. Allow the seeds to settle for another 12 hours. Note the increase in volume due to the mucilaginous action.

ALOE

BOTANY – *Aloe barbadensis* is but one of about 150 species of Aloe. *Barbadensis* is a lily-like succulent-leafed shrub which either has no stem or a 25 cm stem with about 25 leaves in an upright dense rosette. The leaf is thick and fleshy 40-50 cm long, 6-7 cm wide at the base, and lanceolate. The inflorescence is forked once or twice and 60-90 cm high. The raceme is dense cylindrical and narrows toward the top. The flowers are yellow, orange or red, and 3 cm long. The medicinal parts are the dried and fresh juice of the leaves and also the roots. The Aloe plants are indigenous to Africa, cultivated in the Mediterranean region, the Near East, and Asia.

CHEMISTRY – Compounds that are physiologically active include anthracene derivatives particularly anthrone-10-C-glykosyls and 1,8 dihydroxyanthroquinone plus some flavenoids.

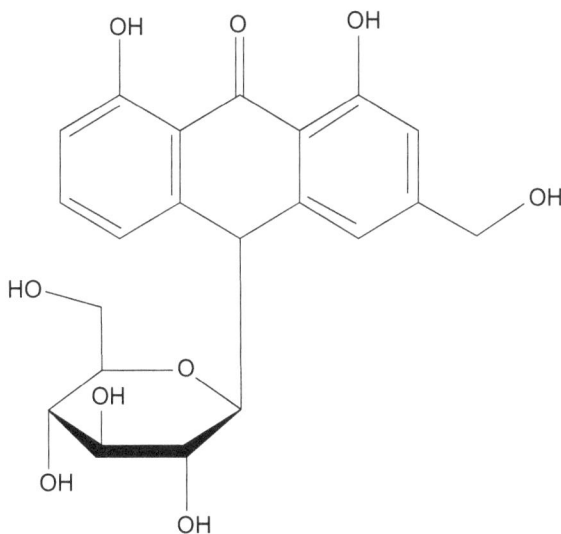

Figure 85. Aloin molecule

Test: Prepare a solution by macerating 1 gm of powdered Aloe in 100 ml of water for 2 hours using a magnetic stirrer agitation. Filter and use the filtrate for the following tests. Mix 5 ml of the filtrate with 45 ml of water and 2 ml of a 5% sodium borate solution to the mixture. The greenish florescence shows the presence of an aloe-emodin anthranol. This is called Schoenteten's Reaction.

Borntraeger's Test: Dilute 10 ml of the filtrate with 100 ml of water and shake the dilution with 10 ml of benzene. Separate the benzene layer and shake it with ammonia TS. A deep rose color should be produced.

Test: Add 2ml of nitric acid to the 5 ml of the filtrate follow with a drop of acetic acid and shake. The result is either a yellow color, a deep red, or a reddish-brown color changing eventually to green depending upon which species of Aloe was tested.

SENNA

BOTANY – Referred to botanically as *Cassia species*, the genus Cassia comprises shrubs, subshrubs, and herbaceous perennials. It has paired-pinnate leaves. The flowers are yellow, occasionally white or pink, and are axillary or terminal in ones, twos, or threes in erect racemes. There are axes with stem glands either between the leaflets or on the petiole. The seeds are numerous and either horizontally or vertically compressed. The habitat of the *Cassia species* is the tropical and subtropical regions of all continents except Europe. Most varieties are indigenous to North, Central, and arid South America. The medicinal part is the leaf.

CHEMISTRY – Composed of anthracene derivatives including senosides A,A1 and B. It also contains some naphthalene derivatives including tinnevelin-6-glucosides.

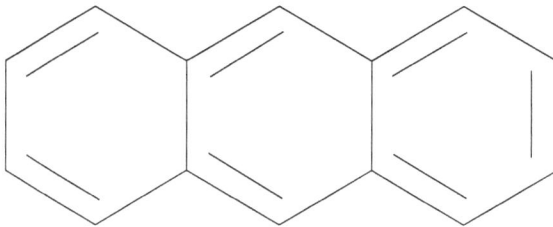

Figure 86. Anthracene molecule

Test: Place 5 gm of Senna in 10 ml of alcoholic potassium hydroxide and boil the mixture for 2 minutes. Dilute with 10 ml of water and filter. Acidify the filtrate with dilute hydrochloric acid and transfer to a small separatory funnel. Shake for a few seconds and allow to stand. Add an equal volume of ether and allow to stand. The ether which contains the emodins will form a separate layer above the aqueous layer. The lower level should then be drawn off and discarded.

Next add 5 ml of dilute ammonium hydroxide to the ether and agitate the mixture again. This action allows the immiscible 1iquids to separate.

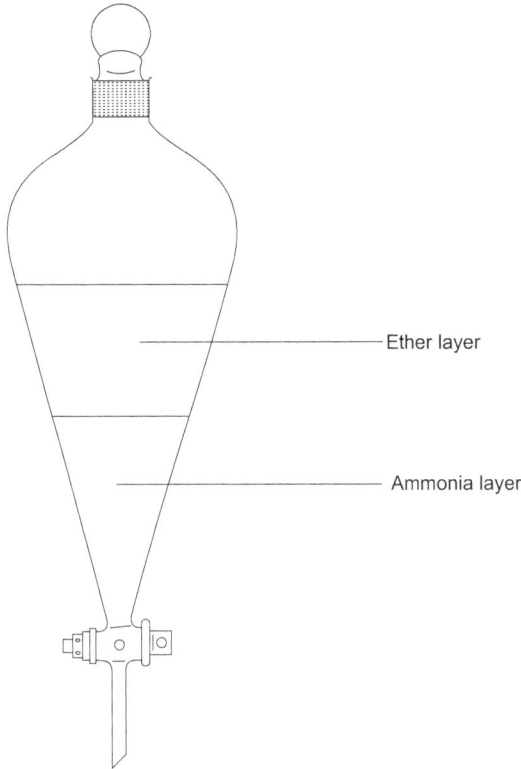

Figure 87. Separatory Funnel with Layers

The lower ammoniacal layer containing the emodins or anthracene will be colored red. This is a positive test for indicating the presence of emodins in senna.

CASCARA SAGRADA

BOTANY – *Rhamnus purshiana*, also called *Frangula purshiana*, are the botanical terms for sacred bark of Cascara Sagrada. The plant is a 6-18 meters tall tree with branches that are gray tomentose when young. The leaves are oblong-ovate rounded at the base or sometimes narrowing at the petiole. The flowers are axillary richly blossomed racemenes. The receptacles are green with larger petals and sepals both of which are white. The medicinal part of Cascara is, as the name suggests, the dried bark. The plant is indigenous to the western part of North America and is cultivated on the Pacific Coast of the United States, Canada, and eastern Africa.

CHEMISTRY – The chief components are anthracene derivatives, particularly 1,8 dihydroxy anthracenes including A and B (stereoisomeric aloin-8-g1ucosides) and C and D (stereoisomeric 11-deoxy-aloin-8-g1ucosides) plus E and F (c-glucosyl-emodin-anthrone-8-glucosides).

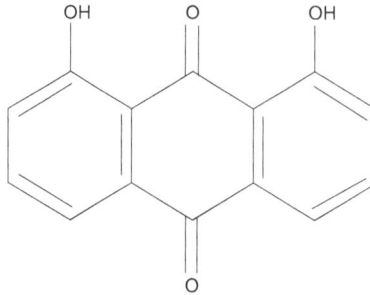

Figure 88. 1,8 Dihydroxy Anthracene

<u>Test for Emodins</u>: Boil 1 gm of bark in 25 ml of sulfuric acid and 100 ml of chloroform in a flask with a reflux condenser. After cooling, transfer to a separatory funnel and draw off the heavier chloroform layer in which the emodins have dissolved from the aqueous layer.

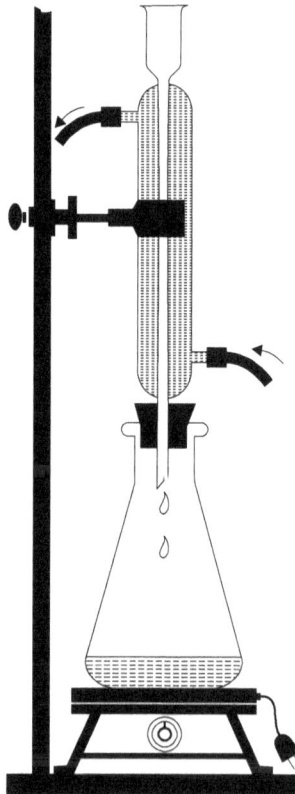

Figure 89. Flask with Reflux Condenser
(**Source:** *A Sourcebook for the Physical Sciences*, Harcourt, 1961, p. 292)

The chloroformic solution is now evaporated to a volume of 10 ml. The solution is transferred to a separatory funnel and extracted with a 5% solution of potassium hydroxide which is diluted to 1 liter of solution. The color of this liquid is now compared with that of a solution produced by dissolving .01 gm of pure emodin in 5% solution of potassium hydroxide and diluting to a volume of 1 liter. Through the process of "Nesslerizing," or color matching by trial and error, one can determine approximately the amount of emodin in the sample of Cascara Sagrada bark.

Additional tests for Cascara Sagrada: Add 0.1 gm of Cascara powder to 10 ml of hot water and shake the mixture occasionally until cool. Filter and dilute the filtrate with sufficient water to make 10 ml, and then add 10 ml of ammonia TS. An orange-yellow color is produced.

Macerate 0.1 gm of Cascara powder with 10 drops of Ethanol. Add 10 ml of water and boil the mixture. Cool, filter, and shake the filtrate with 10 ml of ether. A yellow solution separates.

RHUBARB

BOTANY – *Rheum palmatum* is a large sturdy herbaceous perennial with a stem that grows to over 1.5 meters high. The leaves are orbicularcordate, palmate lobed, somewhat rough on the upper side or smooth, and 3-5 ribbed. The root system consists of a tuber which after a number of years measures 10-15 cm in diameter and has arm-thick lateral roots. The medicinal parts of Rhubarb are the dried underground parts. The plant is indigenous to the western and northwestern provinces of China but is also cultivated in many other parts of the world.

CHEMISTRY – Rhubarb is composed of 108-B glucosides of the aglcone rheumemodin, aloe-emodin, rhein, chrysophanol, and physcion. It also contains tannins, flavonoids, and naphthohydroquinone glycosides.

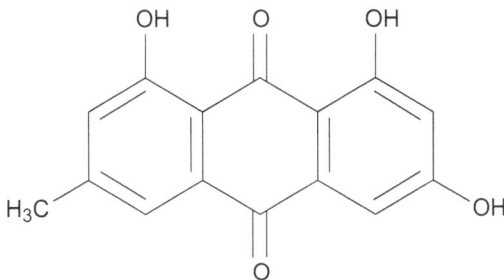

Figure 90. Emodin molecule

Identity Test: Boil 1 gm of rhubarb powder in 10 ml of potassium hydroxide solution. Cool and filter. Then acidify with hydrochloric acid. Shake the mixture with 10 ml of ether. This will produce an ethereal layer that is yellow. Shake the etheral layer with 5 ml of ammonia TS which produces a cherry red color to indicate the presence of emodins.

Tincture of Rhubarb: Prepare by adding equal amounts of Rhubarb, cardamom, and coriander with a 60% ethanol solution. Add some glycerin to smooth the mixture.

GAMBOGE

BOTANY – Gamboge, termed *Garcinia hanburyi* in botanical language, is a tree which grows to about 15 meters high and has a diameter of about 30 cm. The bark is usually in the form of cylindrical sticks that are orange-brown and opaque. The medicinal part of the plant is the resin extracted from the plant. The tree is indigenous to Indochina and Sri Lanka.

CHEMISTRY – Gamboge is composed of 70-75% resins which are mainly yellow or red colored benzophenones and xanthones including Chrysophanic Acid.

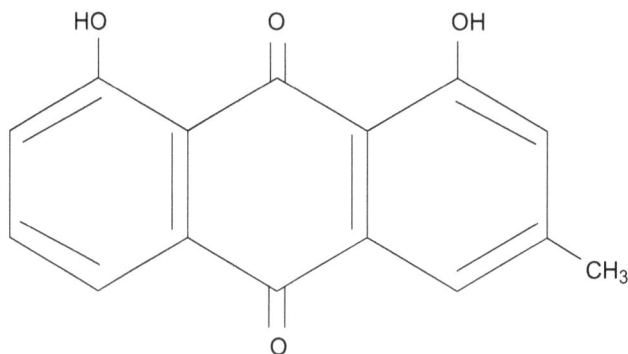

Figure 91. Chrysophanic Acid molecule

Identity Test: Triturate 1 gm of Gamboge powder with water. An emulsion will result having a strong yellow color. Add a few drops of ammonia TS and the mixture becomes dark reddish which fades to clear.

Alcohol-soluble Extractive Test: Macerate about 2 gm of powdered Gamboge with 70 ml of ethanol in a flask. Shake mixture at 30-minute intervals of 8 hours, then allow to stand for 16 hours without shaking. Filter and wash the flask and residue with small portions of ethanol, passing the washings through the filter until the filtrate measures 100 ml. Evaporate 50 ml of the filtrate in a tared dish on a water bath. Dry at 105 °C for 4 hours, cool, and weigh. The weight of the alcohol-soluble extractive is equal to not less than 65% of the weight of the sample of Gamboge used.

ADDITIONAL botanical materials with laxative, cathartic, or purgative properties include Manna, Great Birdweed, Bryonia, Castor Oil, Tragacanth, and Buckthorn Bark.

NEURO-PSYCH

PHARMACY – Neurological and Psychotropic drugs cover a wide range of physiological and neural activities including the autonomic nervous system with its two main divisions: the sympathetic (thoracolumbar) and the parasympathetic (craniosacral). Most organs or systems receive innervation from both of these divisions, but the two can be qualitatively opposed in their action on a given system. The divisions are sometimes referred to as adrenergic and cholinergic. The psychotropic classification includes those drugs which alter the mind and behavior including tranquillizing agents, psychomotor stimulants, and psychotomimetric agents.

MARIJUANA

BOTANY – *Cannabis sativa* may be an annual or a biennial plant which is usually branched and grows up to five meters. It has erect and rough-haired with compressed bristles, long-petioled leaves that are 3-7 pinnate. Hemp is dioecious. The female flowers are reduced to the perigone with one bract and the male form panicles rich in pollen. The medicinal parts are the leaves and the resin exuded from the flowers. The plant probably originated in the Middle East but now is cultivated worldwide.

CHEMISTRY – The chief agent is 9-tetrahydrocannabinol (THC), but there are also other volatile oils and flavonoids.

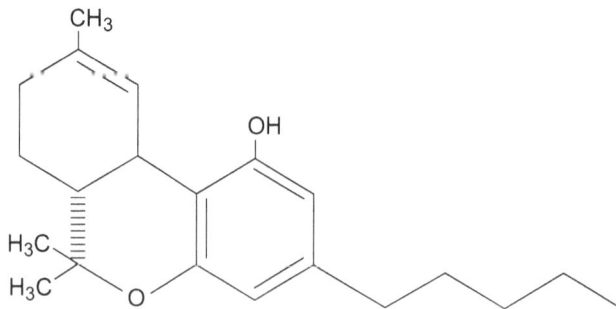

Figure 92. 9-Tetrahydrocannabinol molecule

<u>Preparation</u>: Prepare an extract of Cannabis by percolating 100 gm of Cannabis using ethanol as the menstruum. Macerate the drug for 48 hours and the percolate at a moderate rate until the drug is exhausted. Next, evaporate the percolate at a temperature not exceeding 70 °**C.**

<u>Acid-insoluble Ash</u>: Use the procedure for determining the amount of acid-insoluble ash in Cannabis leaves. The result should show there is less than 5% of the weight of the original sample that remains insoluble in acid.

<u>Isolation of Edestin from Hemp</u>: Begin by thoroughly grinding some Hemp seed in a mortar, then transfer to a small flask with a layer of benzene. Stopper the flask and place on a magnetic stirrer overnight. Filter this by gravity, dispose of the filtrate then repeat the procedure with the seeds. Allow the solvent to evaporate on a watch glass. To 10 gm of the fat-free hemp seeds, add 100 ml of a 1% sodium chloride solution and place in a 250 ml flask. Place in a water bath at 60 °**C** for an hour. Filter the solution until you have collected about 75 ml of filtrate. Warm this turbid solution to about 60 °**C** until the solution becomes clear. Once the solution is clear, discontinue heating and allow the flask to sit in the water bath overnight. This allows for a slow cooling process which results in the formation of larger crystals of edestin than would be formed if the cooling were done more rapidly. Gravity filter the edestin crystals, dry, and weigh. Edestin is classified as a globulin and constitutes about 80% of the proteins of hemp seed which contain the neurological and psychoactive ingredient of Cannabis.

PILOCARPINE

BOTANY – *Pilocarpine microphyllus*, sometime called Jaborandi, is a tree or shrub 3-7 meters high with a trunk diameter of 3-7.5 cm. The branches are pubescent when young and glabrous when older. The leaves are alternate to opposite, odd pinnate with 1-5 pairs of pinna. The numerous flowers are terminal or axillary racemes that are up to 30 cm long and about 0.5 cm wide. The medicinal parts are the dried leaves. The native habitat of the plant is the northeastern part of Brazil.

CHEMISTRY – Pilocarpine is the chief alkaloid of the plant. It is an imidazole type of alkaloid and comprises about 1% of the plant. There are also some companion alkaloids including pilocarpidin and pilosin.

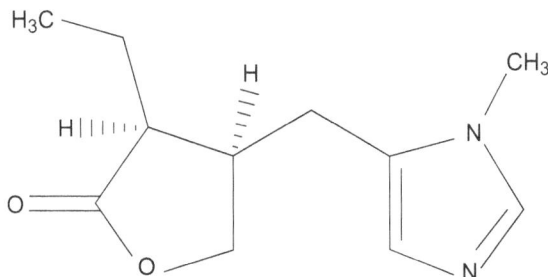

Figure 93. Pilocarpine molecule

Identity Test: Dissolve 10 mg of Pilocarpine hydrochloride in 2 ml of distilled water in a test tube. Add 2 ml of a faintly acid hydrogen peroxide TS and cover the mixture with 1 ml of benzene. Add four drops of potassium dichromate solution (0.3 %) and shake the mixture gently. The benzene layer (top) acquires a violet color while the lower aqueous layer remains yellow.

Test: Mix an aqueous solution of Pilocarpine nitrate with an equal volume of ferrous sulfate TS and superimpose the mixture of 5 ml of sulfuric acid contained in a test tube. The zone of contact becomes brown.

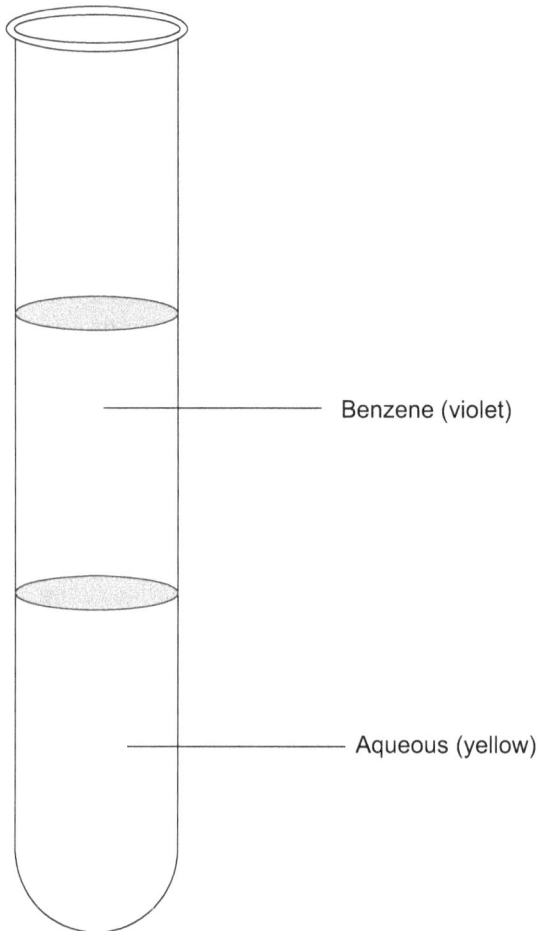

Benzene (violet)

Aqueous (yellow)

Figure 94. Test Tube with Solution Layers

Purity Test: Add some ammonia TS to an aqueous solution of Pilocarpine nitrate. If no turbidity develops in the solution, it indicates that the Pilocarpine is pure and free from any foreign matter.

PEYOTE

BOTANY – *Lophophora williamsii* is the botanical term for Peyote. The plant is a succulent, spineless globular or top-shaped bluish-green cactus with up to 13 distinct vertical ribs. It grows to 20 cm. The roots are tuberous 8-11 cm long. The aerial part has a diameter of 4-12 cm and the pressed-in top is filled with gray wooly bushels of hair. The flowers grow from the center of the cactus head. They are 1-2.5 cm long and 1-2.2 cm across. The outer petals are green with a darker middle stripe and have green-pink or white margins. The medicinal parts are the pin cushion-like aerial transversely cut and dried and the fresh plant. The plant grows in northern Mexico and southern Texas.

CHEMISTRY – The main components are alkaloids, chief of which is Mescaline (up to 7%) but also including hordenine and tetrahydroisoquinoline.

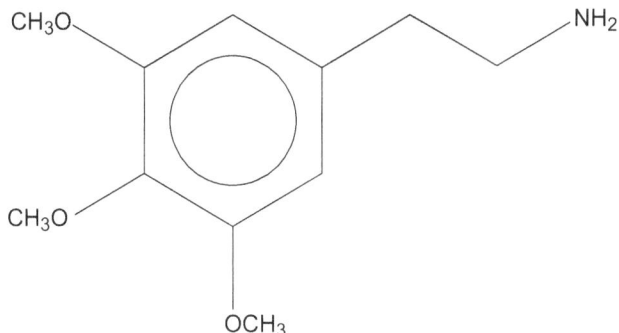

Figure 95. Mescaline molecule

Isolation of Mescaline: Boil 10 gm of peyote in 100 ml of water for 30 minutes. Filter, save the filtrate, and repeat the process. Discard the pulp and boil down to a creamy consistency. Add 4 gm of sodium hydroxide in a separatory funnel along with 20 ml of benzene. Shake, let stand for 2 hours, and benzene layer comes to the top. Draw off the bottom layer, add 10 drops of a sulfuric acid-water combination (2:1) to the benzene layer, and shake thoroughly. Add 10 more drops and crystals of mescaline sulfate appear. Pour the mixture into a funnel and rinse with ether to remove the crystals. Place the mixture and a beaker and put in a heating bath to evaporate the solvents. Dissolve all crystals with 100 ml. Add a few drops of 10 % ammonium hydroxide to bring the pH to 6.5-7. Reduce to 25 ml on a water bath and then place the filtrate in a freezer for 30 minutes. The crystals of Mescaline will gather at the bottom of the beaker. Weigh your product. The yield should be about 7% of the original weight of the Peyote material.

BETEL NUT

BOTANY – *Araeca catechu* is an erect palm tree that can reach 30 meters high. It can have a trunk girth of 50 cm. It has numerous feathery leaflets which are 30-60 cm long, confluent, and glabrous. The flowers are on a branching spandix. The seeds are conical or nearly spherical and about 25 cm in diameter. They are very hard and contain a deep brown testa showing fawn marbling. The medicinal part, as the popular title of the tree suggests, is the nut. The tree is found in the East Indies and also cultivated in parts of Asia and eastern Africa.

CHEMISTRY – Areca is composed of pyridine alkaloids including arecoline, guvacoline, arecaidine, and guvacine. It also has tannins of the catechin type.

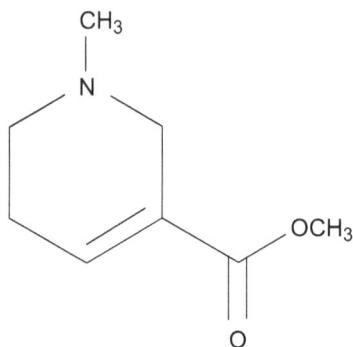

Figure 96. Arecoline molecule

Assay: Place 8 gm of areca powder in a 250 ml flask, then add 80 ml of ether and shake. Add 4 ml of ammonia TS and put on a magnetic stirrer for 10 minutes. Then add 10 gm of anhydrous sodium sulfate and stir for 5 minutes. Allow to settle and decant the ether layer into another flask. Add 0.5 gm of talc to the decanted ether solution and distill off about 2/3rds of the ether. Extract the remaining ether into a separatory funnel with 15 ml of with .02 N sulfuric acid and wash with water using 5 ml of water 3 times. To the combined acid and washings, add methyl red TS and titrate the excess acid with .02 N sodium hydroxide solution using titration burets. Calculate each ml of the .02 N sulfuric acid used is equivalent to 0.0031 gm of Arecoline. The alkaloid content of the Betel Nut is about 20% dominated by the presence of Arecoline.

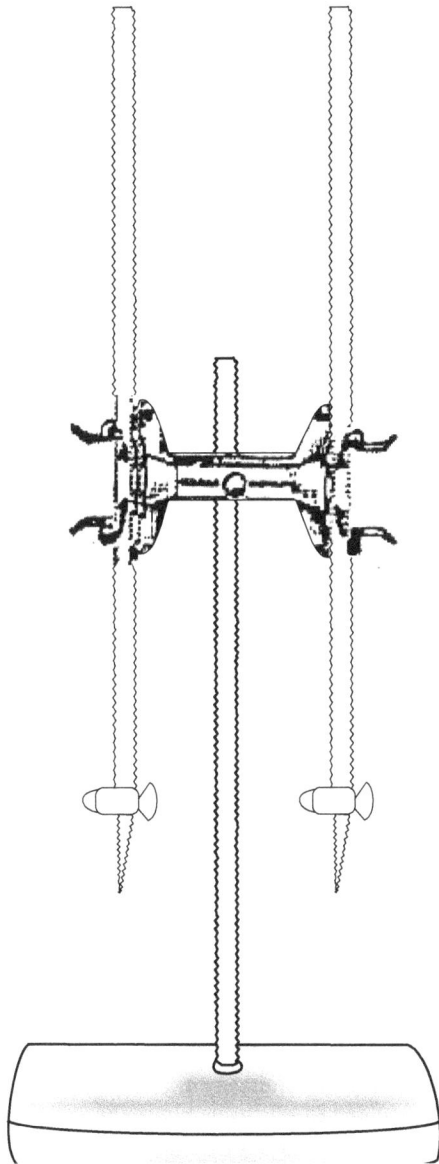

Figure 97. Titration Apparatus
(**Source:** *Chemical Technician's Ready Reference Handbook*, 3rd ed. 1990, p. 619)

COFFEE

BOTANY – The botanical name for Coffee is *Coffea Arabica* which is an evergreen shrub or small tree growing up to 8 meters high with many basal branches. The young branches are glabrous and flattened and nodes which produce many shoots. The bark of the fruiting branches is ashy white. The leaves live 2-3 years and are 6-20 cm long and 2.5-6 cm wide. They are glabrous, slightly coriaceous, dark green, glossy, elliptoid-lanceolate with a distinct leaf tip. The medicinal parts of the plant are the seeds in various forms and stages. Coffee is indigenous to Ethiopia but is cultivated in many tropical regions.

CHEMISTRY – Coffee is composed of purine alkaloids, mainly Caffeine (1-2%) but also trigonelline, 3-5% tannins, 10-13% fatty oils, and also 10-13% of proteins.

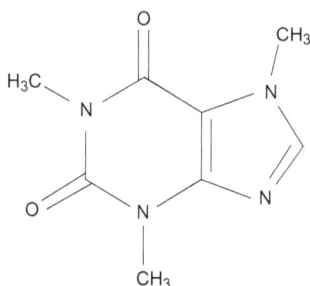

Figure 98. Caffeine molecule

Isolating Caffeine from Coffee: Place 50 gm of Coffee and 250 ml of water in a 600 ml beaker and boil for 15 minutes. Strain through muslin and add a solution of lead acetate to the filtrate until no more precipitate occurs. Heat the mixture to boiling and filter through a Buechner funnel and vacuum system. Heat the filtrate and add dilute sulfuric acid until no more precipitate occurs. Approximately 1 gm of decolorizing charcoal is then added to the mixture which is boiled for a few minutes and then filtered. Collect the filtrate in a large evaporating dish. Add another gm of decolorizing charcoal and the solution evaporated to 100 ml. Cool the mixture, transfer to a separatory funnel, and shake out with 3 successive portions of chloroform. Transfer the combined extracts to a small evaporating dish and warm to dryness. Scrape out the residue, transfer to a small beaker, and dissolve in a small amount of hot ethanol. Let this stand overnight and then filter off the Caffeine crystals whose weight should be 1-2% of the original coffee powder used at the beginning of the procedure.

STRYCHNINE

BOTANY – The *Strychnos nux vomica* tree grows up to 25 meters high with a circumference up to 3 meters. The branches are obtuse-quadrangular, close together, repeatedly bifurcated, glabrous, and have 1 or 2 leaf pairs. The flowers have a 5-tipped calyx and a greenish-white plate-shaped corolla with a long tube. The medicinal parts are the ripe dried seed and the dried bark.

CHEMISTRY – *Nux vomica* contains indole alkaloids, chiefly strychnine and brucine. There are also fatty oils, polysaccharides, and iridoide monoterpenes.

Test: An aqueous solution of strychnine, when introduced to a tube of diphenylamine TS yields a blue color.

Test: a solution of strychnine nitrate, when mixed with a solution of dilute hydrochloric acid, yields a bright red color.

Both of the above are identify tests for strychnine.

Isolation: Mix 8 gm of *Nux vomica* powder with 25 ml of chloroform, 50 ml of ether, and 5 ml of 10% ammonia water. Shake and filter through cotton into a separatory funnel. Add dilute sulfuric acid, then shake and draw off the lower aqueous layer. Render this mixture alkaline by adding some ammonium hydroxide. Completely extract the alkaloidal base by shaking well with several portions of chloroform. Drive off the solvents by heating in an evaporating dish. Once evaporation occurs, add a little ethanol to the residue, dry at 100 °C, and weigh. You have isolated the two major alkaloids of *Nux vomica*—strychnine and brucine.

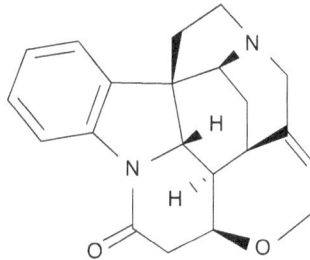

Figure 99. Strychnine molecule

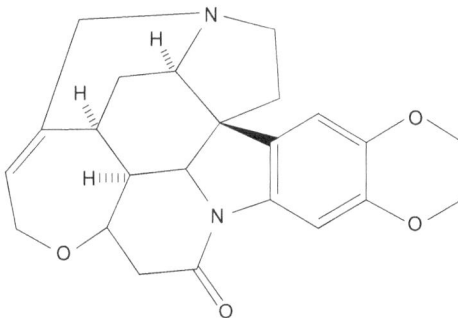

Figure 100. Brucine molecule

ADDITIONAL botanicals with neurological and psychotropic effects include Gotu Kola, Kava-Kava, Ephedra, and Passiflora.

RESPIRATORY

PHARMACY – Respiratory is a term covering a wide range of biological activity. Respiration is controlled by a respiratory center in the medulla oblongata. This center is stimulated by the presence of carbon dioxide in the blood. Respiration is partly controlled by both chemical and sensory stimuli from the carotid body and the carotid sinus. Also respiration may be modified voluntarily through the higher brain centers and drugs which restore the functional activity of depressed higher centers. Drugs with respiratory-affecting properties include respiratory stimulants, expectorants, and antitussives.

EUCALYPTUS

BOTANY – *Eucalyptus globulus* is a deciduous tree growing up to 40 meters high with silver-gray bark which has scattered warts. The trunk is twisted. The juvenile leaves are 7-16 cm by 4-9 cm ovate to broadly lanceolate, cordate, and very glaucus. The mature leaves are 10-13 cm by 3-4 cm lanceolate to fulcate lanceolate, accuminate, asymmetrical, rounded, and glossy green. The flowers are solitary on short petioles. The medicinal part of the tree is the oil extracted from the leaves. Eucalyptus is indigenous to Australia and Tanzania but also cultivated in other tropical areas of the world.

CHEMISTRY – Eucalyptus is composed of volatile oils, mainly eucalyptol (also called cineol) plus other volatile oils including myrtenol and alpha and beta pinenes as well as gamma terpenes and aliphatic aldehydes.

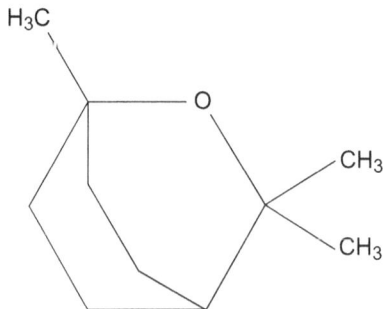

Figure 101. Eucalyptol molecule

Density Test: Weigh and empty 50 ml beaker, then place 10 ml or Eucalyptol in the beaker and weigh again. Repeat the above procedure using water instead of the oil. The relative weight or the oil as the numerator and water as the denominator provides the relative density (specific gravity) of Eucalyptol. Calculation should indicate a density of 0.905-0.925 gm/ml for the Eucalyptol.

Identify Test: Shake 1 ml of Eucalyptol with 20 ml of distilled water and allow the liquids to separate. To 10 ml of the aqueous layer, add one drop or ferric chloride TS. No violet color appears indicating it does not have a phenolic ring.

CAMPHOR

BOTANY – *Cinamomum camphora* is an evergreen tree growing up to 50 meters tell and 5 meters in diameter. The tree is erect at the lower part and knottily branched above. The leaves are 5-11 cm long by 5 cm across, oval lanceolate, alternate, aciminate, grooved, glassy light yellowish-green above, and paler below. The flowers are small and white and are 1-5 mm long pedicles. The fruit is a purplish-black, l-seeded, 10-12 mm oval dupe. The medicinal part is the camphor oil extracted from the tree. The habitat of camphor runs from Vietnam to southern China and as far east as southern Japan.

CHEMISTRY – Camphor is a single substance D (-) camphor, lR,4R 1,7,7 trimethyl-bicyclo (2,2) hepta-2-one.

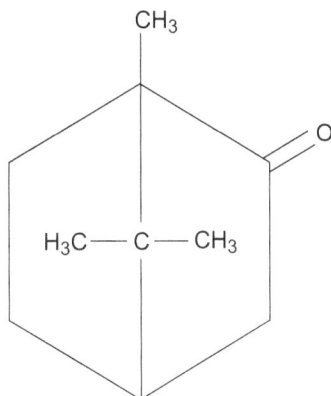

Figure 102. Camphor molecule

Purity Test: Mix 0.1 gm of finely-divided camphor with 0.2 gm of sodium peroxide in a glass test tube. Suspend the tube at an angle of 45° by means of a clamp and heat the tube from the top down to the bottom until incineration is complete. Dissolve the residue in 25 ml of distilled water, acidify with nitric acid, and filter the solution into a comparison tube. Wash the test tube and filter with 2 portions each of 10 ml hot distilled water. Add to the filtrate 0.5 ml of

0.1 M silver nitrate solution, then dilute with distilled water to 50 ml and mix thoroughly. The turbidity is no greater than that produced in a control test with the same quantities of the same reagents and 0.5 ml of .02 M hydrochloric acid.

Preparation: Prepare some Camphor Ointment by melting 22 gm of white wax and 56 gm of lard on a water bath. Then dissolve 22 gm of camphor powder in the melted mixture without further heating and stir the ointment until it is congealed.

CREOSOTE

BOTANY – *Betula species*, or Birch, can grow up to be a 30-meter high tree with a snowy-white bark which usually peels off in horizontal strips or changes into a black, stony-hard bark. Young branches are glabrous and thickly covered in warty resin glands. The petioled leaves are dark green above a lighter gray-green below and have serrate margins and particularly tightly-packed veins. The male flowers are sessile, oblong-cylindrical, and 6-10 cm long. The female catkins are petioled with cylindrical and 2-4 cm long by 8-10 mm thick when fully grown. The medicinal parts are the leaves and the sap resins. Birch is indigenous to Europe from the northern Mediterranean regions to Siberia and to temperate regions of Asia.

CHEMISTRY – Birch compounds include triterpene alcohol esters with saponin-like effects including betula-triterpene saponins. There are also flavonoids including hypertiside, quercetin, myricetin digalactosides plus procenthocyanicidns, volatile oils, and monoterpene glucosides including betula albosides A and B and caffeic acid derivatives.

Identity Test: Add 1 drop of ferric chloride TS to 10 ml of a saturated aqueous solution of Creosote. A violet-blue color develops which is very transient. The liquid becomes cloudy and its color changes rapidly from grayish-green to muddy brown with the final formation of a brown precipitate.

Purity Test: A 2 ml portion of Creosote requires not less than 10 ml and not more than 18 ml of

1.0 M sodium hydroxide to produce a clear liquid as a test for hydrocarbons and bases.

Purity Test: Mix 4 ml of Creosote and 4 ml of glycerin in a test tube and add 1 ml of distilled water. Mix gently and allow to stand. The volume of the creosote layer which separates, equals, or excels the volume of Creosote taken for the test (the so-called "coal-tar creosote" test).

Shake gently 1 ml of Creosote with 2 ml of benzene and 2 ml of freshly-prepared barium hydroxide TS until a uniform mixture results. Upon standing the liquid separates into 3 distinct layers the upper layer being neither blue nor muddy.

ORRIS ROOT

BOTANY – Known botanically as *Iris species*, the Orris Root plant is a perennial 30-100 cm high. The rhizome is thick and short. The flowers are long pedicled and perfumed. The sepals are white or slightly blue. The anthers are as big as the filaments. The medicinal part of the plant is the rhizome with the roots. Orris is indigenous to southern Europe but is cultivated in other regions of temperate climate.

CHEMISTRY – The chief constituent of Orris is a ketone called Irone, in particular alpha, beta, and gamma Irones which give off an odor that resembles the odor of violets. There are also triterpenes present chiefly mono and bispirocyclics plus a number of isoflavonoids including irilon, irisolone, irigenie and tectoridine. Also Orris has small amounts of xanthones and starch.

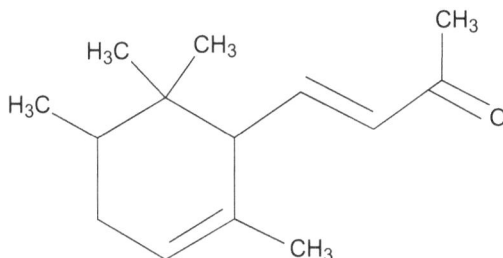

Figure 103. Irone molecule

EPHEDRA

BOTANY – *Ephedra sinica*, also called Ma-Huang, is a 30 cm high lightly-branched subshrub with lengthened cylindrical branches 1-2 mm in diameter. It is similar in appearance to Horsetail and is sometimes twining and often has underground runners. Very small leaves are sometimes reduced to pinted scales and almost fused at the base to form a sheath. They are reddish brown. The medicinal parts are the young canes collected in autumn and the dried rhizome with roots. Ephedra grows mainly in Mongolia and the bordering area of China. *Ephedra gerardiana* is from India.

CHEMISTRY - Ephedra is composed mainly of alkaloids of the 2-aminophenylpropane type including L-ephedrine, IR2S ephedrine, D-pseudoephedrine, L-norephedrine and D-norpseudoephedrine.

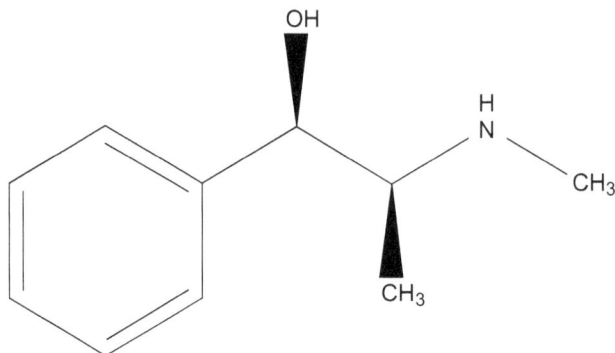

Figure 104. Ephedrine molecule

Moisture Content Test: Using a boiling flask, reflux condenser, and graduated receiving tube, set up a toluene distillation apparatus. Place 10 gm of Ephedra powder in the flask and cover with 200 ml of toluene. Connect the apparatus and fill the receiving tube with toluene poured down through the top of the condenser.

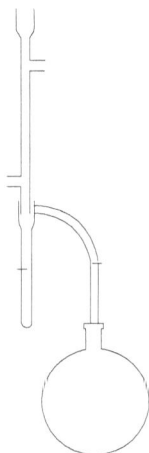

Figure 105. Moisture Content Apparatus
(**Source:** *US Pharmacopeia*, 17th ed, 1960, p. 940)

Heat the flask gently for 15 minutes, and when the toluene begins to boil, distill at the rate of 2 drops per second until most of the water has passed. When complete, rinse the inside of the condenser with a tube brush attached to a copper wire saturated with toluene When the toluene and the water have separated, completely read the volume of water in the graduated receiving tube. Water will be the lower layer since the density of toluene is 0.0866 gm/ml and therefore will rise to the top. Ephedra is highly hygroscopic and can contain a significant amount of water.

LOBELIA

BOTANY – *Lobelia inflata* is an annual or biennial herb 30-60 cm high. The stem is pubescent, angular, branching near the top and contains an acrid latex. The leaves are pale green or yellowish, the lower ones are petiolate the upper ones sessile. The flowers are on long pedicles in the leaf axis. They are pale violet blue and lightly tinged with pale yellow. The medicinal parts are the fresh and dried herbs and the seeds. The plant is indigenous to regions in the north of the United States, in Canada, and Kamchatka, Russia.

CHEMISTRY – The components are chiefly piperidine type alkaloids, particularly lobeline plus companion alkaloids including lobelanine, lobelanidine, norlobelanine, and isolobinine.

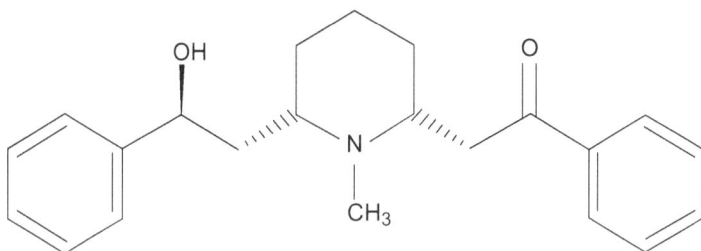

Figure 106. Lobeline molecule

<u>Assay</u>: To 10 gm of lobelia powder mixed with 10 gm of ignited sand, add 75 ml of a mixed solvent of ether-ethanol (4:1) and macerate for 15 minutes. Then add 5 ml of dilute ammonium hydroxide. This mixture should be shaken at intervals for 1 hour. After this step, transfer to a percolator and repeat the procedure with ether-ethanol. The percolate is then transferred to a separatory funnel and extracted first with 30 ml of 0.5 N sulfuric acid and 5 ml of 95% ethanol and finally with 3 further quantities using 20 ml of acid-alcohol mixture. Next, wash the mixture with 3 successive quantities of 5 ml of chloroform, each chloroformic solution being washed with 20 ml portions of 0.5 N sulfuric acid contained in another separatory funnel. The solution is then neutralized with dilute ammonium hydroxide. The alkaloids are then extracted by shaking with 10 ml of chloroform washed with distilled water. The solution is then distilled until about 2 ml of the liquid remains. Then 2 ml of 100% ethanol and the evaporation competed. The mixture is dried at 80 °**C.** The dry alkaloidal material is dissolved in 2 ml of 95% ethanol, 10 ml of .02 N sulfuric acid is added, and the excess acid is titrated using methyl red (range 4.4 red to 6.2 yellow) indicator. Each ml of the acid used is equivalent to 0.00674 gm of alkaloid calculated as lobeline.

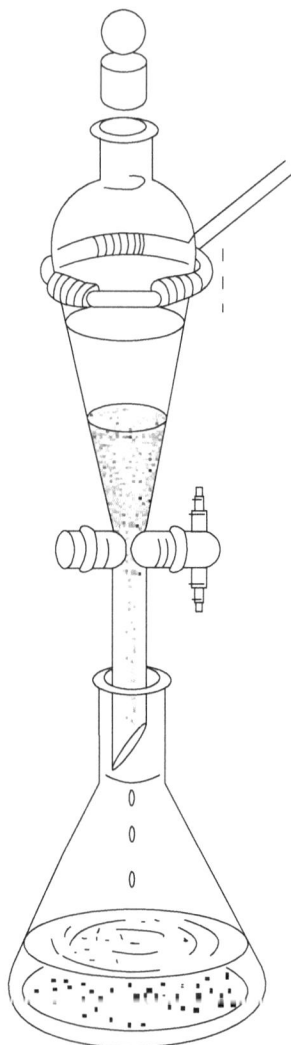

Figure 107. Separatory Funnel with Flask
(**Source:** *Chemical Technician's Ready Reference Handbook*, 1990, p. 439)

ADDITIONAL botanical substances which act upon phases of the respiratory system include poplar bird, white pine, cubeb, licorice, guaiac, turpentine, and quebrache from the aspidosperma.

SEDATIVES

PHARMACY – The term Sedative refers to a quieting effect caused by relaxation and rest but not necessarily sleep. Sedative drugs are used to allay excitement and reduce motor activity with inducing sleep. Sedative and hypnotic qualities sometimes exist in the same drug, a large dose of which may act as a hypnotic, whereas a smaller dose of the same drug may act as a sedative.

RESERPINE

BOTANY – *Rauwolfia sepentina* is an erect glabrous evergreen semi shrub 0.5-1 meter in height. The trunk is pale and unbranched. The leaves are concentrated toward the top of the tree and are entire-margined in whorls of 3-5 and occasionally opposite. The rhizome is vertical and woody. The root is gray-brown with a wrinkled surface 22-32 mm in diameter. The white to pink flowers are in terminal axillary cymes which have a diameter of 2.5-5 cm and are 5-13 cm long. The medicinal part is the dried root. The plant is indigenous to India, Indochina, Borneo, Sri Lanka, and Sumatra.

CHEMISTRY – Rauwolfia is composed primarily of indole alkaloids including reserpine, rescinnamine, sepentinine, raubasine raupone, and ajmaline.

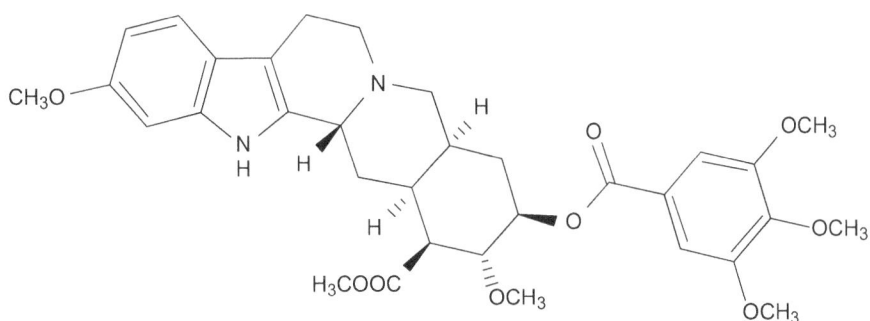

Figure 108. Reserpine molecule

Identity: The UV absorption spectrum of a 1 in 50,000 solution of Reserpine exhibits the same maxima in the range of 230 mu to 350 mu as that of the USP Reserpine Reference similarly measured and the respective absorptivities determined with reference to a mixture 3.6 volumes of chloroform and 1.4 volumes of methanol as the blank at the points of maximum absorbance occurring at about 268 mu and 295 mu do not differ more than 3%.

Identity: To 0.5 ml of glacial acetic acid and one drop of the assay solution, add 1 ml of a 1 in 50 solution of vanillin in hydrochloric acid. A pink color is produced and becomes deep violet-red within a few minutes as result of warming the solution 10-20 seconds.

HYOSCYAMUS

BOTANY – *Hyoscyamus niger* or Henbane is an erect herb with simple leaves. The root is fusiform and turnip-like at the top. The leaves are oblong, roughly crenate-dentate and gray-green. The basal leaves are petiolate and the cauline leaves are stem-clasping. The medicinal parts are the dried leaves and seeds and the whole fresh flowering plant. Henbane is indigenous to Europe and western and northern Asia and northern Africa. It has also been introduced to eastern Asia, North America, and Australia.

CHEMISTRY – The compounds are mainly tropane alkaloids including the chief one—hyoscyamine—and to a lesser extent, atropine, and scopalamine.

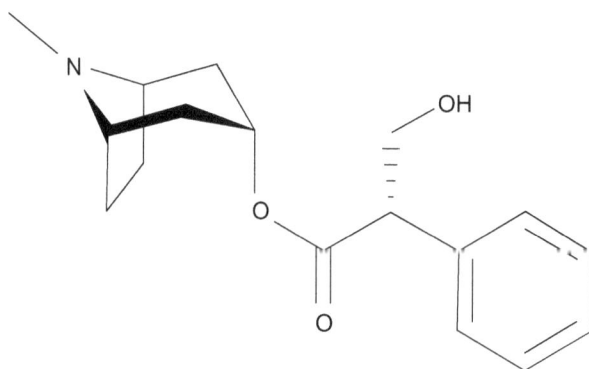

Figure 109. Hyoscyamine molecule

Identity: A 1% solution of hyoscyamine mixed with a few drops of platinum hexachloride yields a buff color as a characteristic reaction of a tropane alkaloid.

ST. JOHN'S WORT

BOTANY – *Hypericum perforatum*, called St. John's Wort, is a perennial with a long-living, fusiform, branched root and branched rhizome. The stem is erect with two raised edges, reddish-tinged, and 100 cm high. The leaves are oval-oblong, entire-margined, opposite, translucent, punctate, sessile, and often covered in black glands. The golden yellow flowers are in sparsely blossomed terminal cymes. The medicinal parts include the fresh buds, the aerial parts collected during the flowering season, and the whole fresh flowering plant. St. John's Wort is indigenous to all of Europe, western Asia, and northern Africa. It has been introduced to eastern Asia, Australia, and New Zealand, and is cultivated in other parts of the world.

CHEMISTRY – Compounds in St. John's Wort include some anthracene derivatives favoring naphthadihydrodianthrones, in particular hypericin and pseudo-hypericin. There are also flavonoids chiefly hyperoside, quercitrin, rutin, and isoquercitrin and also bioflavonoids including, among others, amentoflavone plus xanthones (1,3,6,7 tetrahydroxy-xanthone) and acylphloroglucinols hyperiform with small quantities of adhyperiform volatile oils including aliphatic hydrocarbons such as 2-methyl octane, undecane and dodecanol and sesquiterpenes including, among others, alphapinenes, carophyllene and 2-methyl-3-but-en-2-o1and oligomeric procyanidines, catechin tannins and caffeic acid derivatives including chlorogenic acid.

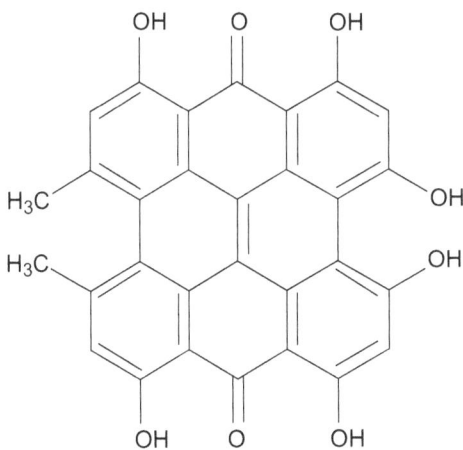

Figure 110. Hypericin molecule

Identity: Place 0.5 gm of Hypericin in 5 ml of pyridine. Observe the cherry-red florescence

Identity: Place 0.5 gm of Hypericin in alkaline solution with pH ranging between 8 and 11. Note the red color of the solution.

Identity: Place 0.5 gm of Hypericin in an alkaline solution of pH 12 or higher. A green solution is the result with a red florescence.

BITTER ORANGE

BOTANY – *Citrus aurantium* is an evergreen tree with a rounded crown and smooth grayish-brown bark. The branches are angular when young, becoming terete and glabrous soon after with axillary spines. The flowers are arranged singly or in clusters in the axis and are very fragrant. The peel is thick, rough, and orange when ripe. The plant is indigenous to Asia but cultivated in other tropical and subtropical regions of the world. The medicinal part is the fresh and dried fruit peel.

CHEMISTRY – Components of Bitter Orange include volatile oils such as linalool, linalyll acetate, alphapinenes, limonene, nerol, and geraniol. Also there are flavonoids including hesperidin, dihydrochalcone, and naringin as well as lipophilic compounds including sinensetin, nobelitin, and tangeretin.

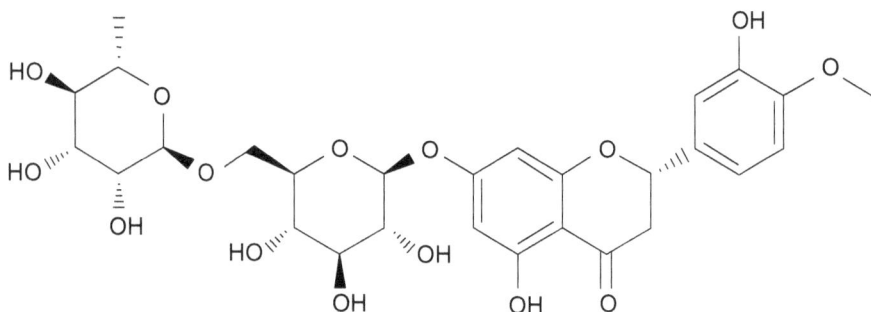

Figure 111. Hesperidin molecule

Isolation of Hesperidin from Orange Peel: Besperidin is the major flavonoid in orange and lemon peel and can be extracted successively with ether and methanol, the first solvent removing the essential oil and the second, the glycoside. Place 200 gm of dried orange peel in a round-bottomed flask that is attached to a reflux condenser. Add 1 liter of ether and heat on a water bath for an hour. Filter the contents of the flask through a Büchner funnel and allow the powder to dry at room temperature. Then return the dry powder to a flask, add 1 liter of methanol, and heat under reflux for 3 hours. This leaves a syrupy residue which may then be crystallized from dilute acetic acid yielding white needles with a melting point 252-254 °C.

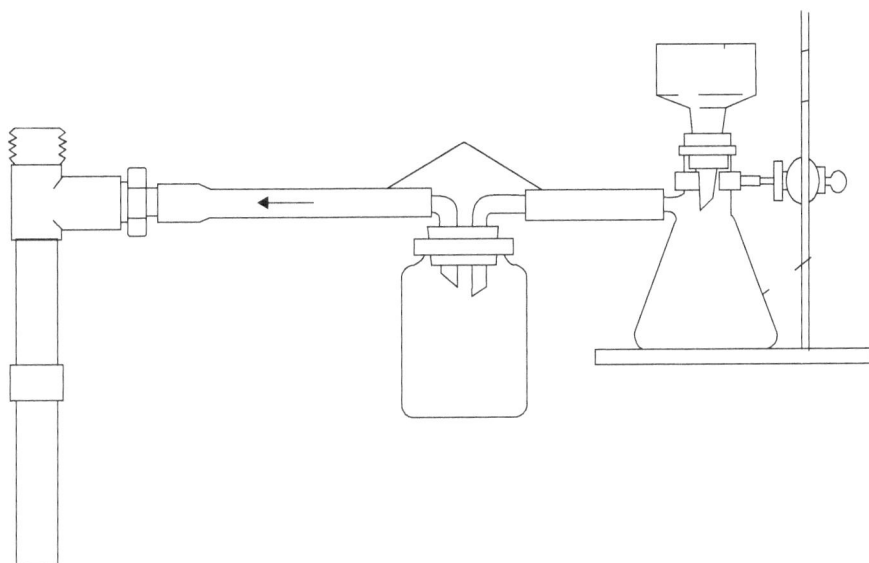

Figure 112. Vacuum Filtration Apparatus

<u>Purifying the Hesperidin</u>: Use a 10% solution of the drug in formamide prepared by warming to 60 °**C** and treating with activated charcoal that has been treated with dilute hydrochloric acid. Filter through celite diluted with an equal volume of water and allow to stand for a few hours to give time to crystallize. The crystals of the purified Hesperidin are then filtered off, washed with hot water and next with isolpropanol yielding a white crystalline product.

WOOD BETONY

BOTANY – *Betonica officinalis* is a plant growing a height of 30-100 cm. The stem is erect, unbranched, quadrangular, bristly-haired, and usually has only 2 distal pairs of leaves. The basal leaves are rosette-like. The leaves are elongate-ovate with a cordate base and crenate. The lower ones are larger and long-petioled, and the upper ones are smaller and shorter. The medicinal parts of the herb are the basal leaf. The plant is native to Europe.

CHEMISTRY – The chief component of Wood Betony is Betaine including betonicine (-) oxystachydrine, stachydrine and (-) oxystachydrine. There are also caffeic acid derivatives including chlorogenic acid isochlorogenic acid and rosemary acid iridoid glycosides.

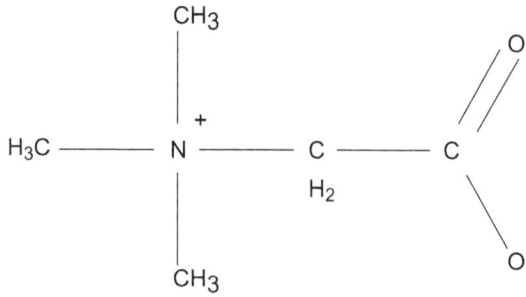

Figure 113. Betaine molecule

ADDITIONAL botanicals with sedative properties include horehound, feverfew, linden, and Jamaica dogwood.

SKELETAL-MUSCULAR

PHARMACY – The Skeletal Muscular system is obviously the largest body system in terms of area. Skeletal muscle may be relaxed by blocking the effects of somatic motor nerve impulses or by depressing the appropriate neurons within the central nervous system so that the somatic nerve impulses fail to be generated. This category of drug actions include those drugs which act at the myoneural junction, called neuromuscular blocking drugs, and those drugs that act upon central neurons or centrally-acting muscle relaxants.

ASH

BOTANY – Ash, known botanically as *Fraxinus excelsior*, is an impressive 15-30 meter high tree with gray-brown, smooth, later fissured and wrinkled bark and large black-brown pubescent buds. The leaves are entire margined, opposite and odd pinnate usually 5-11 cm long by 1-3 cm wide, oblong-ovate, long acuminate, finely and sharply serrate. The flowers are on richly blossomed panicles, the terminal ones appearing on the new flower branches. The medicinal parts are the dried leaves, the fresh bark, the branch bark, and the dried leaves. The plant is distributed in most of Europe except the northern, southern and eastern edges.

CHEMISTRY – Tree components are made of the flavonoid, Rutin plus tannins, mucilages, iridoide monterpenes, and hydroxucoumarins including aesculin and fraxin.

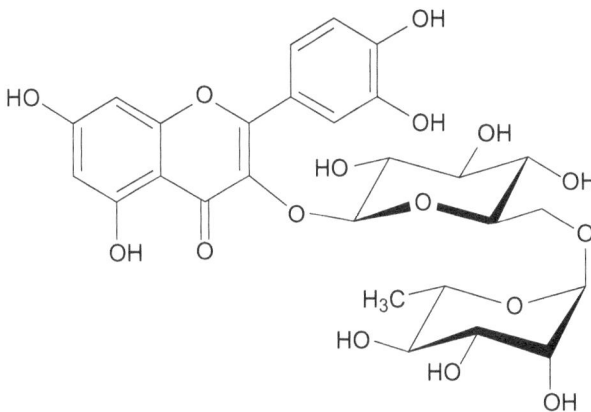

Figure 114. Rutin molecule

Test: Add 5 ml of a rutin/hydrochloric acid solution filtrate to a test tube and neutralize the excess acid with sodium hydroxide TS. Add 3 ml of hot alkaline cupric tartrate TS and suspend in a water bath for 10 minutes. A red precipitate of cuprous oxide results.

Test: Dissolve 10 mg of Rutin in 10 ml of ethanol, then add 1 drop of ferric chloride TS. In a few minutes a greenish-brown color results.

Test: Add a few crystals of Rutin to 5 ml of sodium hydroxide TS. An orange-yellow color will develop as another identity test for Rutin.

SPEEDWELL

BOTANY – Speedwell, known botanically as Veronica officinalis, is a 10-20 cm high herbaceous perennial with runners that tend to form grass. The root system consists mainly of shoot-producing roots. The stem is creeping, and the flower-bearing branches are erect. The medicinal parts are the dried herb collected during flowering season. The plant is indigenous to almost all of Europe, parts of Asia, and North America. Drug sources are mainly from Bulgaria, Hungary, and parts of the former Yugoslavia.

CHEMISTRY – Speedwell is composed of iridoide monoterpenes including Aucubin and numerous catalpol esters including minecoside, verminoside, veronica coside, mussaenoside and labroside. There are also flavonoids including luteolin 7-0 glucosides and triterpenes, saponins, plus caffeic acid derivatives such as chlorogenic acid.

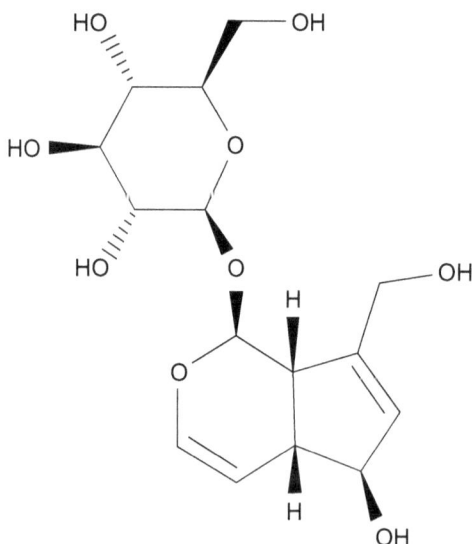

Figure 115. Aucubin molecule

ROSE HIP

BOTANY – *Rosa canina* is a plant approximately 1-3 meters high climbing and trailing prickly shrub with erect root shoots covered in sickle-shaped prickles. The leaves are pinnatifid with 5-7 leaflets which are petiolate elliptical to ovate, serrate, glabrous, and dark green above lighter beneath. The long pedicled white or pale pink flowers are usually solitary or in clusters of 2 or 3. The medicinal parts are the petals and the hips with and without seeds. *Rosa canina* grows in Europe and northern Africa and is extensively cultivated in many regions of the world.

CHEMISTRY – The main compounds is Ascorbic Acid (Vitamin C) but there are also other fruit acids including malic and citric plus pectins, sugars, tannins, carotinoids, flavonoids, volatile oils, and proteins.

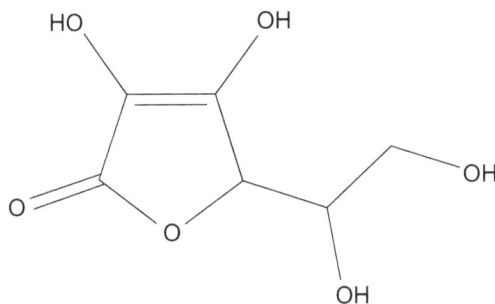

Figure 116. Ascorbic Acid molecule

Molecular Weight of Ascorbic Acid: Place 100 ml of water in 250 ml boiling flask and heat gently and record the temperature (should be 100 °C). Discontinue heating, cool, and then add 5 gm of Ascorbic Acid. Reheat gently until the boiling point of the solution is reached.

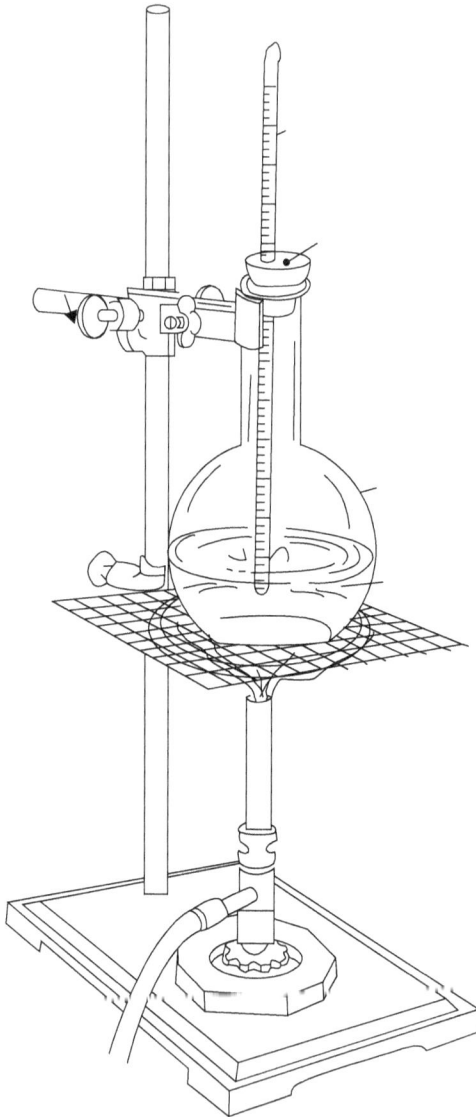

Figure 117. Molecular Weight Apparatus
(**Source:** *Chemical Technician's Ready Reference Handbook*, 3rd ed., McGraw-Hill, 1990, p. 431)

Calculate the difference which is the ΔT_b value. Enter the number of grams of solute per kilogram of solvent (water), K_b of 0.52 as the boiling constant for water. Using this formula, calculate the molecular weight of the Ascorbic Acid, which should be about 176.12 gm.

Formula: $\underline{K_b \text{ x gm (solute)}}$
                ΔT_b x kg (solvent) = molecular weight

BLACKCURRANT

BOTANY – *Ribes nigra* is a sturdy perennial bush up to 2 meters high. The branches are pale, hard, and initially pubescent. The leaves are alternate, petiolate, becoming quickly glabrous on the upper surface with numerous yellow resin glands on the undersurface. The flowers form richly blossomed racemes. Each is in the axil of a pubescent bract, which is shorter than the petiole. The medicinal parts are the leaves collected and dried after the flowering season. Plant habitat is the Eurasian forest as far as the Himalayas plus Canada and Australia.

CHEMISTRY – Blackcurrant is composed of flavonoids including astragalus, rutin, and isoquercitin plus fruit acids including malic and citric in addition to sugars and pectins.

Test: Place a few drops of ferric chloride TS with a solution of astragalus (tragacanth), boil briefly, and note the stringy yellow precipitate.

Test: A stringy precipitate is also formed when a boiling solution of astragalus is mixed with a solution of copper hydroxide in 20% ammonia.

BLACK MUSTARD

BOTANY – *Brassica nigra* is an annual tall-growing, slim, branched plant with fusiform roots. The stem grows up to 1 meter and is almost round and bristly-haired at the base with a bluish bloom toward the top. The medicinal part is the seed. The plant is distributed worldwide.

CHEMISTRY – The main active compound of Black Mustard is the volatile oil allyl isothiocyanate.

Formula: $S=C=N-CH_2-CH=CH_2$

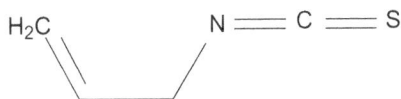

Figure 118. Allyl isothiocyanate molecule

Identity: A small amount of powdered Black Mustard is mascerated in water for 2 hours and then distilled. One drop of the distillate will give fine needle in a few ml of phenylhydrazine.

Test: Dilute the distillate with one-half of its volume with ammonia water TS. Allow to stand for a few hours and permit a drop to evaporate on a slide. Crystals of thiosinamine are formed which vary considerably depending on concentration and the presence of impurities.

Test: Add a drop of silver nitrate TS to this thiosinamine solution and observe the long needle mostly branching or in bundles or rosettes being formed.

POISON IVY

BOTANY – *Rhus toxicodendron* is the botanical name for the well-known Poison Ivy. The plant is a dioecious shrub which grows up to one meter high with ascending, procumbent, or climbing rooting branches and underground runners. The branches are initially green and softly pubescent but later they become brown and glabrous. The leaves are trifoliate with 8-14 cm long petioles. The leaflets are oblong acute or obtuse, entire margined or roughly serrate in the middle. The pedicled flowers are in axillary pubescent panicles. They are dioecious, sometimes androgynous. The stem petals are whitish-green with red hearts. The fruit is almost globular, glabrous, yellow or yellowish-white, and 10-grooved drupe. The fruit varies in size and contains a viscous latex in resin channels which turns black in the air. The medicinal parts are the leaves collected after flowering and dried, the young flowering branches, and the fresh leaves. Poison Ivy is native to North America but is also found in East Asia and is cultivated in Germany in botanical and apothecary gardens.

CHEMISTRY – The chief compound of Poison Ivy is the phenol named Urushiol in such forms as cis-cis 3 (n-heptadeca 8,11 dienyl) catechol; cis-cis cis-3 (n-heptaca 8,11,14 trieneyl) catechol; and cis-3 (n-heptadec-8-enyl) catechol. In addition there are also some tannins and flavonoids.

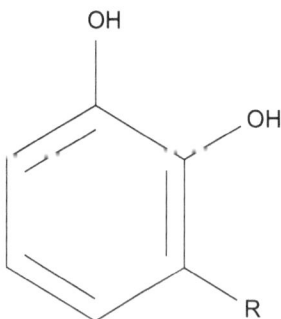

Figure 119. Urushiol molecule

1. $R=(CH_2)_{14}CH_3$
2. $R=(CH_2)_7CH=CH(CH_3)_5CH_3$
3. $R=(CH_2)_7CH=CH_2CH=CH(CH_2)_2CH_3$
4. $R=(CH_2)_7CH=CHCHCH_2CH=CHCH=CHCH_3$

5. $R=(CH_2)_7CH=CHCH_2CH=CHCH_2CH=CH_2$

Test (Phenolic): Carefully place 2ml of Urushiol in a test tube and add a few drops of ferric chloride TS. Note the violet color which appears as the evidence of the presence of a phenol.

Test: Carefully place 2 ml of Urushiol in a test tube and a few drops of bromine TS. A white precipitate appears which dissolves first but becomes permanent as more of the reagent is added.

VALERIAN

BOTANY – *Valerian officinalis* is a plant that grows 50-100 cm high. It has a short cylindrical rhizome with finger-length bushy round roots. The stem is erect and unbranched. The leaves are odd-pinnate with 11-23 lanceolate undented-dentate leaflets. The androgynous flowers are in panicled cymes. The medicinal parts are the carefully dried underground parts and the dried roots. The plant is found in Europe and in the temperate regions of Asia. It is also cultivated in England, Japan, and the United States.

CHEMISTRY – The many compounds of Valerian include isovaleric acid, is-ovaleroxyhydroxy, didavaltrate, and acevaltrate. Volatile oils include isovaleric acid, isoeugenyl valernate, and cryptofaurinol. There are also various sesquiter-penes and pyridine alkaloids.

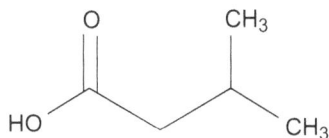

Figure 120. Isovaleric Acid molecule

Fluid Extract of Valerian preparation: Macerate 5 gm of valerian powder in a solution of ethanol: water (4:1) for a period of 48 hours. Percolate this at a moderate rate.

Tincture of Valerian preparation: Mix the ground valerian powder with the menstruum. Use 5 gm of the valerian and about 3 ml of an ethanol:water (3:1) to render the drug evenly and distinctly damp. Now transfer the dampened drug to a suitable percolator and allow it to stand for 15 minutes. Then pack it firmly and pour on a sufficient quantity of menstruum above it to cover the top of the percolator. When the liquid is about to drop from the percolator close the lower orifice and allow the drug to mascerate for 3 hours adding fresh menstruum as needed. After the 3-hour period, percolate the tincture at a moderate rate.

ADDITIONAL botanicals which affect the skeletal and muscular areas are turmeric, ginger, thuja, cayenne, cajaput, vervain, verbena, horse chestnut, and devil's claw.

FORMULAS FOR TEST SOLUTIONS

Alkaline Cupric Tartrate TS: Dissolve 34.06 gm of cupric sulfate in sufficient water to make 500 ml. Store this solution in tight containers. Dissolve 173 gm of potassium sodium tartrate and 50 gm of sodium hydroxide in sufficient water to make 500 ml. Store this solution in small alkali-resistant containers. For use, mix exactly the same volumes of the two solutions at the time required.

Bromine Water TS: Agitate 2 ml of bromine with 100 ml of cold water in a glass-stoppered bottle. The stopper of the bottle should be lubricated with petrolatum. Store in a cold place protected from light.

Bromophenol Blue TS: Dissolve 100 mg of bromophemol blue in 100 ml of diluted alcohol. Filter if necessary.

Bromothymol Blue TS: Dissolve 100 mg of bromothymol blue in 100 ml of diluted alcohol. Filter if necessary.

Diphenylamine TS: Dissolve 1.0 gm of diphenylamine in 100 ml of sulfuric acid. The solution is colorless.

Ferric Ammonium Sulfate TS: Dissolve 8 gm of ferric ammonium sulfate in sufficient water to make 100 ml

Ferrous Ammonium Sulfate TS: Dissolve 9 gm of ferrous ammonium sulfate in sufficient water to make 100 ml.

Ferrous Sulfate TS: Dissolve 8 gm of ferrous sulfate in 100 ml of recently boiled and thoroughly cooled water.

Hydroxylamine Hydrochloride TS: Dissolve 3.5 gm of hydroxylamine hydrochloride in 95 ml of 60% ethanol and add 0.5 ml of bromophenol blue solution until a greenish tint develops.

Add sufficient ethanol to make 100 ml.

Iodine TS: Dissolve 14 gm of iodine in a solution of potassium iodide (30 gm in 100 ml of water). Add three drop of hydrochloric acid.

Phenolphthalein TS: Dissolve 1.0 gm of phenolphthalein in 100 ml of alcohol.

Potassium Dichromate TS: Dissolve 7.5 gm of potassium dichromate in 100 ml of water.

Potassium Ferricyanide TS: Dissolve 1.0 gm of potassium ferricyanide in 10 ml of water. Prepare this solution fresh.

Potassium Ferrocyanide TS: Dissolve 1.0 gm of potassium ferrocyanide in 10 ml of water. Prepare this solution fresh.

Potassium Hydroxide (Alcoholic) TS: Dissolve 35 gm potassium hydroxide in 20 ml of water and add sufficient alcohol to make 1000 ml. Store in tightly stoppered bottles and protect from light.

Silver Nitrate TS: Dissolve 17.5 gm of silver nitrate in 1000 ml of water. Store in as dark a place as possible.

Thymol Blue TS: Dissolve 100 mg of thymol blue in l00 ml of alcohol.

BIBLIOGRAPHY:

Barton, D. *Comprehensive Natural Products Chemistry*, Pergamon, 1988

Bean, H. A. *Advances in Pharmaceutical Sciences*, Academic, 1982

Beckett, H. *Practical Pharmaceutical Chemistry*, Humanities, 1988

Bennett, P. *Clinical Pharmacology*, Elsevier, 2012

Briggs, T. *Biochemistry*, Springer-Verlag, 1991

Dickson, C. *Medicinal Chemistry Laboratory Manual*, CRC, 1988

Fiorkin, K. *Analytical Profiles of Drug Substances*, Academic, 1981

Gleason, F. *Plant Biochemistry*, Jones & Bartlett, 2011

Gokhals, S. *Pharmacognosy*, Pragamatic, 2008

Handbook of Chemistry and Physics, 92nd ed., CRC, 2012

Heinrich, M. *Fundamentals of Pharmacognosy*, Elsevier, 2011

Ikan, R. *Natural Products*, Academic, 1991

Kapoor, L. *Handbook of Ayurvedic Medicine*, CRC, 2000

Makowski, G. *Advances in Clinical Chemistry*, Elsevier, 2012

Martin, A. *Physical Pharmacy*, Williams & Wilkins, 1995

The Merck Index, 14th ed., Merck, 2011

Nelson, C. *Principles of Biochemistry*, Freeman, 2008

Patrick, G. *Introduction to Medicinal Chemistry*, Oxford UP, 2009

Physicians' Desk Reference for Herbal Medicines, Medical Economics, 1998

Remington's Science and Practice of Pharmacy, Lippincott, 2008

Robbers, J. *Pharmacognosy Laboratory Guide*, Purdue UP, 1981

Shugar, G. *Chemical Technicians' Ready Reference Handbook*, McGraw-Hill, 1990

Tyler, V. *Pharmacognosy*, Lea & Febiger, 1988

US Pharmacopeia and National Formulary, USPC, 2011

Wagner, H. *Economic and Medicinal Plants*, Academic, 1991

Wigglesworth, J. *Biochemical Research*, Wiley, 1983